# A THEORETICAL STUDY OF INTERPHASE MASS TRANSFER

# A THEORETICAL STUDY OF
# INTERPHASE
# MASS TRANSFER

ROBERT W. SCHRAGE

1953

COLUMBIA UNIVERSITY PRESS, NEW YORK

Copyright 1953 Columbia University Press, New York

Published in Great Britain, Canada, and India by
Geoffrey Cumberlege, Oxford University Press
London, Toronto, and Bombay

Manufactured in the United States of America

## CLARKE F. ANSLEY AWARD

This study was selected by a committee of the Faculty of Pure Science, Columbia University, to receive the Clarke F. Ansley Award for 1951. Three such awards are given by Columbia University Press for outstanding dissertations submitted in candidacy for the degree of Doctor of Philosophy in the three Graduate Faculties of the University.

## ACKNOWLEDGMENT

The author wishes to acknowledge his indebtedness to Professor Thomas B. Drew of Columbia University, whose advice and encouragement were largely responsible for this research being undertaken and accomplished.

## CONTENTS

### Chapter I

| | |
|---|---|
| INTRODUCTION | 3 |
| 1.1. The Problem | 3 |
| 1.2. Some Concepts of the Kinetic Theory of Gases | 7 |
| 1.3. The Nature of a Gas-Liquid or Gas-Solid Phase Interface at Equilibrium | 14 |
| 1.4. Results of Encounters between Gas Molecules and Liquid or Solid Surfaces | 16 |
| 1.5 The Theory of Viscous Slip and Temperature Jump | 20 |

### Chapter II

| | |
|---|---|
| THE ABSOLUTE RATE OF VAPORIZATION OF A PURE SUBSTANCE | 25 |
| 2.1 Systems at Equilibrium | 25 |
| 2.2 Systems at Nonequilibrium | 28 |
| 2.3 Experimental Evidence | 29 |

### Chapter III

| | |
|---|---|
| THE SIMPLE THEORY OF INTERPHASE MASS TRANSFER OF A PURE SUBSTANCE | 32 |
| 3.1 The Absolute Rate of Condensation under Nonequilibrium Conditions | 32 |
| 3.2 Interphase Mass Transfer of a Pure Substance | 36 |
| 3.3 Interfacial Nonequilibrium for the Special Case of $T_0/T_s = 1$ | 38 |
| 3.4 Experimental Studies of Interphase Mass Transfer | 40 |

### Chapter IV

| | |
|---|---|
| INTERPHASE MASS TRANSFER IN MULTICOMPONENT SYSTEMS | 44 |
| 4.1 The Absolute Rate of Vaporization for a Component of a Multicomponent Liquid or Solid | 44 |

## CONTENTS

- 4.2 The Absolute Rate of Condensation for a Component of a Multicomponent Gas under Nonequilibrium Conditions — 46
- 4.3 Interphase Mass Transfer in a Binary System — 48

### Chapter V

**ANOTHER THEORY OF INTERPHASE MASS TRANSFER OF A PURE SUBSTANCE — 50**

- 5.1 The Exact Description of the Gas at the Phase Interface — 50
- 5.2 The Crout Theory of the Evaporation of a Pure Substance — 51
- 5.3 The Assumed Distribution Function for the Gas at the Phase Interface — 52
- 5.4 Transport of Mass, Momentum, and Energy in the Gas at the Phase Interface — 54
- 5.5 Transport of Mass, Momentum, and Energy in the Uniform Gas — 59
- 5.6 Condition of the Uniform Gas at Plane 1 as a Function of the Rate of Interphase Mass Transfer — 60
- 5.7 The Maximum Rate of Interphase Mass Transfer — 62
- 5.8 Comparison of the Various Theories for Interphase Mass Transfer of Monatomic Molecules — 64
- 5.9 Extension of the Theory to Polyatomic Molecules — 67

### Chapter VI

**PREDICTED DEPARTURES FROM INTERFACIAL EQUILIBRIUM IN TYPICAL MASS TRANSFER PROCESSES — 72**

- 6.1 Interphase Mass Transfer at Ordinary Pressures — 72
- 6.2 Interphase Mass Transfer at Reduced Pressures — 74

### Chapter VII

**CONCLUSION — 77**

LIST OF SYMBOLS — 81
APPENDIXES — 83
LIST OF SOURCES CITED — 99

# A THEORETICAL STUDY OF INTERPHASE MASS TRANSFER

# CHAPTER I

## INTRODUCTION

**1.1. The Problem.** The transfer of heat, mass, and momentum between phases is an important part of most engineering operations. It is, in fact, almost impossible to think of any dynamic process in which one or more of these phenomena are not involved. The analysis of these transfer operations can usually be separated into two parts, the more familiar one of which might be called intraphase transfer by diffusional processes. Included within this group are thermal conduction, mass diffusion, and viscous transfer of momentum. An examination of intraphase processes cannot, however, be a complete consideration of the general problem of interphase transfer. There remains the question of how molecules or the properties associated with them are transferred from one phase to another. Strictly speaking, the adjective "interphase" applies only to this consideration, which is the subject of the present study, particularly in so far as it applies to the transfer of mass between phases.

The mechanism of transfer across phase boundaries has not received nearly so much attention as the diffusional transfer processes, because it has usually been felt adequate to assume that nearly complete equilibrium always exists at a phase interface. In the theory of fluid flow, for example, it is usual to presume that, when two phases move past each other, the velocity of one relative to the other is zero at their interface. In heat transfer the interfacial temperatures of the two phases are assumed identical, while in mass transfer the compositions are taken to correspond to their values at thermodynamic equilibrium.

The purpose of this book is to examine carefully the mechanism of interphase mass transfer and to determine under what conditions, if any, the assumption of thermodynamic equilibrium is inadequate. The discussion is limited to gas-liquid and gas-solid systems because these are most susceptible to theoretical examination. This research was undertaken chiefly because of its possible application to the engineering of mass transfer processes. Although it was not originally intended to examine the more fundamental aspects of the problem, consideration from this viewpoint cannot be entirely avoided.

As is fairly well known, the early history of fluid flow theory shows a long period of uncertainty regarding the possibility of "slip" of a fluid past a solid boundary. The inability to explain a large part of the experimental data on fluid flow in terms of the molecular viscosity seemed to suggest the possible occurrence of such an effect. Also, in the early development of heat transfer theory, Poisson mentioned the possibility of a discontinuity in the temperature distribution of a fluid near a solid surface, analogous to the velocity distribution discontinuity already mentioned. The discovery of turbulence in the latter part of the nineteenth century, however, led to a different explanation of the discrepancies which had been noted between early theories and data. This new viewpoint gradually dispelled much of the previous doubt as to conditions just at a phase interface, and it came to be commonly accepted that the velocities or temperatures of two contacting phases were the same at their interface. Further discrepancies between theory and data were, in general, explained by the inadequacy of intraphase theory or by the inaccuracy of the data.

At about this same time, research undertaken to test some of the predictions of the kinetic theory of gases found definite indications of the effects which had long been suspected. In 1875 Kundt and Warburg (46)[1] were engaged in experimental determination of the viscosity of gases at low densities, where real behavior most nearly corresponds to the assumed behavior on which kinetic theory is based. To their surprise, the difference between predicted and experimental viscosities increased rather than diminished when the pressure was reduced beyond a certain point. This anomaly was resolved by the explanation that in these experiments the gas immediately adjacent to a solid surface seemed to have a finite velocity relative to the surface. An analogous effect was predicted for heat transfer processes and was discovered experimentally by Smoluchowski (86) in 1898. Later, we will have occasion to consider these effects in detail, but their discovery received no great attention in engineering circles at that time because their occurrence was limited to low pressure systems which were not of particular interest.

Theories of intraphase mass transfer, or ordinary diffusion, developed later than related work in fluid flow or heat transfer. It was not until 1855 that Fick (25) first suggested the concept of mass diffusivity in analogy to thermal conductivity. In 1873 Stefan (87) measured gas phase diffusivities by a method which involved interphase mass transfer between a gas and a liquid. A narrow tube was partly filled with a liquid so that its vapor would diffuse through the gas space at the top of the tube and disperse into the surrounding atmosphere. If the concentration of vapor at the top of the tube is taken as zero, while that at the phase interface is taken as the saturation concentration corresponding to the

---

[1]Numbers in parentheses refer to the "List of Sources Cited" at the end of the book.

liquid temperature, then the diffusivity of the vapor in the gas can be calculated from the rate of evaporation.

Although Stefan considered that substantially complete equilibrium existed at the phase interface in his experiments, he clearly stated that this might not always be the case:

> Es ist möglich dass bei Versuchen in luftverdünnten Raume, in welchem die Diffusion des Dampfes viel rascher wird, der Druck des Dampfes an der Oberfläche der Flüssigkeit fortwährend kleiner bleibt, als das der Oberflächentemperatur entsprechende Maximum der Spankraft ist.[2]

Stefan went on to suggest that under these circumstances the difference between the equilibrium vapor pressure and the actual partial pressure of the vapor at the interface could be related to the rate of mass transfer by a simple proportionality factor. This factor might be considered a new property of the liquid called the "vaporization coefficient." From this viewpoint, Stefan's own work was to be regarded as a special case in which the actual departure from interfacial equilibrium was negligible.

In 1884 some similar experiments were made by Winkelmann, who observed that, if the length of the gas space at the top of the tube were made greater, causing a lower rate of evaporation, then the diffusivity apparently increased (100). Since the diffusivity was not expected to be a function of the rate of diffusion, Winkelmann offered two alternative explanations. Either the more rapid rates of vaporization caused a nonisothermal condition within the apparatus, so that the temperature of the liquid surface was lower than that of the surrounding constant temperature bath, or, even though temperature uniformity might have existed, there could possibly be an appreciable departure of the vapor concentration at the phase interface from its equilibrium value. Evidently, Winkelmann felt that the latter explanation was the correct one.

In 1910 Mache (56) made further experiments and observed similar results. He also discounted the possibility that this effect might be caused by temperature gradients in the apparatus, since even slow vaporization of water apparently produced larger departures from equilibrium than did rapid vaporization of ether or alcohol. Following Stefan's suggestion, Mache defined a vaporization coefficient, $K_0$, by the equation

$$w = K_0(P_s^* - P_0), \quad (1.1\text{-}1)$$

where $w$ is the rate of mass transfer per unit area, $P_s^*$ is the vapor pressure corresponding to the temperature of the liquid surface, and $P_0$ is the partial pressure of the vapor in the gas at the phase interface.

In 1916 LeBlanc and Wuppermann (51) made modifications in the

[2]See reference 87, pp. 407-408.

design of the diffusion apparatus which were intended to eliminate the possibility of temperature nonuniformities. They observed no variation of the diffusivity with the rate of mass transfer and concluded that the effects noted by Winkelmann and Mache were caused by a cooling of the liquid surface rather than by any true departure from interfacial equilibrium.

At about this same time, there began to appear the so-called film theory for mass transfer in packed columns and in other equipment for contacting phases. In 1916 Lewis (53) presented the first clear statement of this idea, which suggested that, when two phases move past each other, the more fluid one tends to form a stationary film on the surface of the other. The rate of mass transfer was supposedly controlled by molecular diffusion through this film. At its outer boundary, turbulence kept the composition of the fluid more or less uniform, while at the phase interface, the two phases were "probably nearly in equilibrium." Later, the two-film theory of Whitman (98) replaced the single-film concept, but the idea of interfacial equilibrium was retained and commonly accepted.

In 1930, however, Nusselt (67), who was aware of the work of Mache, questioned the adequacy of the assumption of interfacial equilibrium. Acting on a suggestion by Nusselt, Ackermann (1) in 1934 reported that a study of the psychometer by him had definitely established appreciable departures from equilibrium at the phase interface during mass transfer processes. According to Ackermann, the results of this study indicated that (1.1-1) was true only at low rates of mass transfer. At high rates, the difference between the interfacial partial pressure of the vapor and the equilibrium vapor pressure approached a limiting maximum value. The rates of interphase mass transfer ordinarily encountered in engineering operations were said usually to be in this high range, where the departure was constant. Ackermann tabulated values of $\gamma_s^*/\gamma_0$, the ratio of the equilibrium vapor density to the actual vapor density of water in air at the phase interface, as a function of temperature. This information is reproduced in Table I. Unfortunately, no details were given on the work which is the basis of these results.[3] However, both the magnitude and nature of the effect suggested by Ackermann seem very unreasonable.

In 1937 and 1938 Mache (57, 58) published two additional papers on the vaporization coefficient. Here he introduced a new idea by attempting

---

[3] According to a footnote on p. 131 of reference 1, the research on which these results were based was to be reported shortly under the title "Eine neue Theorie des Aspirationspsychometers." A search of several abstracting journals for the following five years failed to reveal any paper by the author under this title.

# INTRODUCTION

### TABLE I

Ackermann's (1) values of the maximum departure from interfacial equilibrium for the evaporation of water into air ($\gamma_s^*/\gamma_0$ = ratio of equilibrium to actual mass density of water at interface).

| $t$, °C | 11 | 20 | 30 | 40 | 46 | 52.2 | 58 |
|---|---|---|---|---|---|---|---|
| $\gamma_s^*/\gamma_0$ | 1.18 | 1.22 | 1.28 | 1.37 | 1.46 | 1.57 | 1.70 |

to measure the surface temperature of the liquid by means of a fine thermocouple and thus eliminate any question as to temperature nonuniformity. It was reported that impurities at a liquid surface interfere considerably with the vaporization process. This interference could be avoided either if care was taken to use pure liquids or if their evaporating surfaces were constantly renewed. At about this same time, Schirmer (79) also attempted to measure directly the surface temperature of a liquid during evaporation. The work of neither of these investigators will be discussed, however, since later experiments by Prüger (71), which will be referred to subsequently, were more refined.

In 1935 Higbie (35) studied the absorption of gases by liquids during short periods of exposure. Because a diffusional theory proposed by him failed to correlate the data obtained, he concluded that the assumption of interfacial equilibrium was incorrect for the conditions of his experimental work. He finally interpreted his data in terms of a "first-order process" at the liquid surface, but no details of the mathematical analysis were given.

The literature cited in this section will serve to give some idea of the nature of the problem before us. Other contributions on the subject of interphase mass transfer may be most conveniently mentioned at appropriate points in the later theoretical development. First, however, some consideration will be given to certain topics which are either closely related or fundamental to the main subject of this study.

**1.2. Some Concepts of the Kinetic Theory of Gases.** The kinetic theory of gases finds application in any theoretical discussion of the nonequilibrium behavior of a gas phase in contact with a liquid or solid. While the present research was not undertaken because of any primary interest in the fundamentals of such behavior, it is necessary to approach the problem from this viewpoint. Some of the fundamental definitions and theorems of kinetic theory are reviewed here so that persons unfamiliar with them need not refer to another source.[4] The

---

[4] Parts of the content of Section 1.2 are discussed in such standard references as Chapman and Cowling (21) and Kennard (39).

following discussion applies to pure gases unless it is specifically stated otherwise.

*(a) The velocity distribution function.* The mass center of a molecule may be described by a point in a system of six coordinates. Three of these locate its position in the ordinary sense and are simply the space or Cartesian coordinates $x$, $y$, and $z$, referred to an arbitrary system of mutually perpendicular axes. The radius vector $r$, which is a function of $x$, $y$, and $z$, is also used to locate the mass center in Cartesian space. The remaining three coordinates, $U$, $V$, and $W$, describe the motion of the mass center by locating it in velocity space. $U$, $V$, and $W$ are the scalar magnitudes of the $x$, $y$, and $z$ components of the velocity $C$ of the mass center. The velocity $C$ is referred to the velocity of the origin of the Cartesian coordinate axes.

The statistical behavior of the mass centers of a system of molecules is ordinarily adequate to describe many observable physical phenomena. The velocity distribution function is defined as a convenient mathematical representation of such behavior. If $dn$ is the number density of molecules which lie in a volume element $dC$ of velocity space at the velocity $C$, then

$$dn = f\, dC, \qquad (1.2\text{-}1)$$

where $f$ is, by definition, the velocity distribution function. For steady-state processes it is, in general, a function of $r$ and $C$. An equivalent scalar form of (1.2-1) would be

$$dn = f\, dU dV dW. \qquad (1.2\text{-}1a)$$

Equation (1.2-1) or (1.2-1a) can be used to calculate any property of a gas which is an additive function of the motion of the mass centers of its molecules. For the sake of brevity, this section will use the vector form of equations, but the equivalent scalar equations should be obvious.

The total value per unit volume of any additive molecular property $\psi$ at a given point may be computed as

$$\sum \psi = \int \psi f\, dC, \qquad (1.2\text{-}2)$$

where the summation $\Sigma$ is made over all the molecules per unit volume at the point and the integration extends over all of velocity space. The property $\psi$ may be a scalar, vector, or tensor property of the molecule. The average value of any molecular property at the point $r$ is

INTRODUCTION 9

$$\bar{\psi} = \frac{1}{n} \int \psi f d\mathbf{C},  \qquad (1.2\text{-}3)$$

where $n$ is the number density of molecules at the point.

If the total value per unit volume of a molecular property is desired for a certain group of molecules within a restricted portion of velocity space between the limits $\mathbf{C}'$ and $\mathbf{C}''$, it may be obtained as

$$\sum_{\mathbf{C}' \to \mathbf{C}''} \psi = \int_{\mathbf{C}'}^{\mathbf{C}''} \psi f d\mathbf{C}. \qquad (1.2\text{-}4)$$

The average value of a molecular property for this restricted group of molecules is

$$\bar{\psi}_{\mathbf{C}' \to \mathbf{C}''} = \frac{\int_{\mathbf{C}'}^{\mathbf{C}''} \psi f d\mathbf{C}}{\int_{\mathbf{C}'}^{\mathbf{C}''} f d\mathbf{C}}. \qquad (1.2\text{-}5)$$

*(b) The uniform gas.* If a gas is in a uniform steady state, so that its macroscopic properties do not vary with time or position, then statistical mechanics predicts the form of the velocity distribution function defined in (1.2-1). Since a consideration of the specific structure of the molecules or of the nature of the collisions between them is not involved, the result has a wide applicability to real systems. It is known as the Maxwell velocity distribution function and is

$$f = A e^{-\beta^2 (\mathbf{C} - \bar{\mathbf{C}})^2} \qquad (1.2\text{-}6)$$

or

$$f = A e^{-\beta^2 [(U - \bar{U})^2 + (V - \bar{V})^2 + (W - \bar{W})^2]}, \qquad (1.2\text{-}6a)$$

where $A$ and $\beta$ are constants which depend upon the thermodynamic state of the gas and $\bar{\mathbf{C}}$ is its linear velocity due to simple mass motion.

Substitution of (1.2-6) in (1.2-1) and integration over all velocity space imposes a necessary condition upon the two constants $A$ and $\beta$, that is,

$$A = n \frac{\beta^3}{\pi^{3/2}}. \qquad (1.2\text{-}7)$$

The constant $\beta$ may be related to the thermodynamic properties of a perfect gas by a consideration of its pressure tensor, as will be shown. The result of this analysis may be anticipated now, however, and is

$$\beta = \sqrt{\frac{M}{2RT}}, \qquad (1.2\text{-}8)$$

where $M$ is the molecular weight, $R$ is the perfect gas constant, and $T$ is the absolute temperature. It will usually be convenient to use the distribution function as

$$f = n\left(\frac{\beta^3}{\pi^{3/2}}\right) e^{-\beta^2(\mathbf{C} - \overline{\mathbf{C}})^2} \qquad (1.2\text{-}9)$$

or

$$f = n\left(\frac{\beta^3}{\pi^{3/2}}\right) e^{-\beta^2[(U - \overline{U})^2 + (V - \overline{V})^2 + (W - \overline{W})^2]}. \qquad (1.2\text{-}9a)$$

If a uniform gas consists of more than one molecular specie, then the equations (1.2-6) through (1.2-9a) are applicable to each individually.

A further result of classical statistical mechanics is the theorem of equipartition of energy. If the expression for the total energy of a molecule consists of a number of terms, $s$, each of which is proportional to the square of a particular variable, then each of these terms (for an average molecule of mass $m$) contributes an amount $m/4\beta^2$ to the total molecular energy. For example, each of the components of the translational velocity (excepting that part due to the motion of the gas as a whole) gives rise to a kinetic energy term which has an average value given by

$$\frac{1}{2}m\overline{(U - \overline{U})^2} = \frac{1}{2}m\overline{(V - \overline{V})^2} = \frac{1}{2}m\overline{(W - \overline{W})^2} = \frac{m}{4\beta^2}. \qquad (1.2\text{-}10)$$

The total energy of an average molecule with velocity $\mathbf{C}$ may, according to classical kinetic theory, then be expressed as

$$\epsilon = \frac{1}{2}m\mathbf{C}^2 + \frac{m}{4\beta^2}(s - 3). \qquad (1.2\text{-}11)$$

The second term of (1.2-11) represents the internal energy of the molecule and is not correlated with $\mathbf{C}$, but has the same average value through all of translatory velocity space.

# INTRODUCTION

The application of quantum theory to statistical mechanics indicates the well-substantiated fact that all of the possible internal energy terms of a molecule may not be fully excited. In this case, particular terms may contribute only a fractional part of $m/4\beta^2$ to the total energy of an average molecule. However, so far as use is made of (1.2-11) in this study (see Section 5.9), $s$ may in this case be regarded as a nonintegral number which is a function of the temperature of the system.

*(c) The nonuniform gas.* Whenever heat, a tangential component of momentum, or a particular molecular specie in a mixture flows steadily through a plane which moves with the linear velocity of a gas, then the gas is no longer in a uniform state. In these cases the theorems of statistical mechanics no longer apply, except that they predict a great tendency for the gas to achieve a uniform state, so that only slight deviations from (1.2-9) are to be expected. These deviations are significant, however, and the exact determination of the transport properties of a gas is impossible unless they are considered.

It can easily be shown that (1.2-9) is inconsistent with heat conduction, diffusion, or viscous flow in gases.[5] Nevertheless, there have been many treatments of these processes without resort to determination of a correct distribution function. When used in an approximate way, (1.2-9) can predict the order of magnitude of the thermal conductivity, diffusivity, and viscosity of gases. Beyond this rough approximation, however, this method of analysis is unsatisfactory, particularly for mass transfer processes. While it may have several degrees of approximation, depending upon various refinements which can be incorporated in it, it is fundamentally an incorrect approach.

The attention of such investigators as Chapman and Enskog accordingly turned to the determination of the correct form of distribution function. The starting point of such derivations is the Boltzmann integro-differential equation:

$$\frac{Df}{D\theta} = \frac{\partial f}{\partial \theta} + \mathbf{C} \cdot \nabla f = \left(\frac{\partial f}{\partial \theta}\right)_{\text{collisions}}, \qquad (1.2\text{-}12)$$

where $\theta$ is time. The right-hand side of this equation can be expressed as an integral function of $f$ which depends specifically on the results of collisions between molecules. The complicated mathematical treatment from this viewpoint is discussed in Chapman and Cowling (21). The results of this analysis are in relatively good agreement with experimental observations. It is also possible to obtain the principal results of this derivation by less elegant methods.[6] Furry (28) has

---

[5] See reference 39, p. 146.
[6] See reference 39, Chapter IV.

pointed out that in certain respects the early Stefan-Maxwell treatment of diffusion is more consistent with the correct formulation of Chapman and Enskog than were some of the later interpretations. Furry also obtains the results of the correct theory for diffusion without considering (1.2-12) at all, by merely applying an approximate correction factor to (1.2-9), which happens to convert it into the first approximation for the distribution function deduced from (1.2-12) by Chapman and Enskog. The choice of the correction factor seems superficially logical, but it was not suggested until long after it had been determined theoretically.

*(d) Pressure effects in pure gases.* The pressure, or force per unit area, exerted by a gas on an impenetrable and nonvolatile surface is equal to the rate at which molecules carry momentum to and from that surface.[7] The concept of pressure within a gas is retained by defining it as the force per unit area which would be exerted on a plane moving with the gas so that there is no net flow of matter across it. The momentum of a molecule relative to such a plane is $m(C - \bar{C})$, and the rate of flow of momentum across the plane is $\mathbf{n} \cdot (C - \bar{C})(C - \bar{C})m$, where $\mathbf{n}$ is a unit normal to the plane.[8] From (1.2-2) this flow of momentum may be integrated over all of velocity space to give the pressure acting on the plane:

$$\mathbf{P}_n = \int \mathbf{n} \cdot m(C - \bar{C})(C - \bar{C}) f dC \,, \qquad (1.2\text{-}13)$$

$$\mathbf{P}_n = \mathbf{n} \cdot \int m(C - \bar{C})(C - \bar{C}) f dC \,. \qquad (1.2\text{-}14)$$

Comparing (1.2-14) with (1.2-3), we see that (1.2-14) may also be written

$$\mathbf{P}_n = \mathbf{n} \cdot [\gamma \overline{(C - \bar{C})(C - \bar{C})}] \,, \qquad (1.2\text{-}15)$$

where $\gamma$ is the mass density of the gas. The bracketed expression in (1.2-15) is, by definition, the pressure tensor, $\mathcal{P}$. If $\mathbf{i}$, $\mathbf{j}$, and $\mathbf{k}$ are unit vectors extending in the directions of the $x$, $y$, and $z$ axes, then the pressure tensor may be written in terms of its component dyads as

$$\mathcal{P} = \gamma \begin{bmatrix} \overline{(U-\bar{U})(U-\bar{U})}\mathbf{ii} + \overline{(U-\bar{U})(V-\bar{V})}\mathbf{ij} + \overline{(U-\bar{U})(W-\bar{W})}\mathbf{ik} \\ + \overline{(V-\bar{V})(U-\bar{U})}\mathbf{ji} + \overline{(V-\bar{V})(V-\bar{V})}\mathbf{jj} + \overline{(V-\bar{V})(W-\bar{W})}\mathbf{jk} \\ + \overline{(W-\bar{W})(U-\bar{U})}\mathbf{ki} + \overline{(W-\bar{W})(V-\bar{V})}\mathbf{kj} + \overline{(W-\bar{W})(W-\bar{W})}\mathbf{kk} \end{bmatrix} . \qquad (1.2\text{-}16)$$

---

[7] Other effects occur in dense gases, but dense gases are of little importance in the subject of this research.

[8] The unit vector $\mathbf{n}$, whose scalar magnitude is unity, is not to be confused with the scalar quantity $n$, which has been defined as the number density of molecules.

INTRODUCTION   13

Since there is no flow of matter through an impenetrable and nonvolatile boundary, the definition of the pressure tensor is consistent with the ordinary concept of the pressure exerted by a gas. If n is a unit normal to such a boundary, then the pressure on the boundary is

$$\mathbf{P}_n = \mathbf{n} \cdot \wp. \tag{1.2-17}$$

It should also be clear that, if molecules penetrate a surface, but the rates of penetration and emission are such as to cause no net flow of matter through the surface, then the pressure tensor continues to give the pressure both in the gas and on the surface.

When interphase mass transfer occurs, however, a further consideration is necessary. Reynolds (75, 76) was evidently the first to note that transfer processes may affect the pressure acting on a surface, but his discussion of the effect was not correct. The pressure tensor defined in (1.2-16) continues to give the conventional pressure in the gas phase, no matter what process may be occurring there. However, if one imagines a plane which moves with the velocity of the liquid or solid surface, $\mathbf{C}_s$, and proceeds to examine the rate at which momentum is carried to and from it, one is led to define a new tensor which will be called the surface pressure tensor, $\wp_s$, and is given by

$$\wp_s = \gamma \begin{bmatrix} \overline{(U-U_s)(U-U_s)}\mathbf{ii} + \overline{(U-U_s)(V-V_s)}\mathbf{ij} + \overline{(U-U_s)(W-W_s)}\mathbf{ik} \\ + \overline{(V-V_s)(U-U_s)}\mathbf{ji} + \overline{(V-V_s)(V-V_s)}\mathbf{jj} + \overline{(V-V_s)(W-W_s)}\mathbf{jk} \\ + \overline{(W-W_s)(U-U_s)}\mathbf{ki} + \overline{(W-W_s)(V-V_s)}\mathbf{kj} + \overline{(W-W_s)(W-W_s)}\mathbf{kk} \end{bmatrix}.$$

$$\tag{1.2-18}$$

If n is a unit normal to the liquid or solid surface, then the pressure acting on the surface is

$$\mathbf{P}_s = \mathbf{n} \cdot \wp_s. \tag{1.2-19}$$

Ordinarily, in any mass transfer process between a liquid or a solid and a gas, the relatively great density of the condensed phase causes $\mathbf{C}_s$ to be small compared with $\overline{\mathbf{C}}$. For consistency with certain other simplifications made later, we shall usually consider $\mathbf{C}_s$, and in particular its component normal to the surface, to be negligible.

The hydrostatic pressure of a gas is conventionally defined as one third the scalar of its pressure tensor. In the case of a uniform gas, it is related to the thermodynamic equation of state for the gas. If the components of the pressure tensor of a uniform gas are evaluated from (1.2-3) and (1.2-9a), it will be found that all but the diagonal dyads in (1.2-16) vanish. The magnitudes of these terms are equal and are

$$\overline{\gamma(U-\bar{U})^2} = \overline{\gamma(V-\bar{V})^2} = \overline{\gamma(W-\bar{W})^2} = \frac{\gamma}{2\beta^2}. \quad (1.2\text{-}20)$$

By definition, the hydrostatic pressure of a uniform gas is then

$$P = \frac{1}{3}\left[\mathbf{i} \cdot \overline{P} \cdot \mathbf{i} + \mathbf{j} \cdot \overline{P} \cdot \mathbf{j} + \mathbf{k} \cdot \overline{P} \cdot \mathbf{k}\right] = \frac{\gamma}{2\beta^2}. \quad (1.2\text{-}21)$$

This can be combined with the perfect gas equation of state to give

$$P = \frac{RT}{M}\gamma = \frac{\gamma}{2\beta^2}, \quad (1.2\text{-}22)$$

from which the value of $\beta$ given by (1.2-8) immediately follows.

**1.3. The Nature of a Gas-Liquid or Gas-Solid Phase Interface at Equilibrium.** Before any theories of mass transfer between a gas and a liquid or solid phase are discussed, it will be desirable to consider the information which is available on the description of a phase interface at equilibrium. Direct observation of the interface by means of a magnifying device does not show other than an abrupt change in the uniform properties of the two phases. However, the limited resolution of such equipment precludes the possibility of examining variations which occur over distances approaching molecular dimensions. It is necessary at this point to have some idea as to whether nonuniformities in the gas phase might exist over distances of the order of a mean free path from the interface. Much of the theory to be developed is based on the presumption that the influence of a liquid or solid surface does not extend into the gas beyond a region which approximates more than a small fraction of a mean free path.

The treatments which have appeared in the literature on interphase mass transfer tacitly make this assumption. The gas and condensed phases are considered to be essentially uniform and to be separated only by a mathematical plane. While the transition between the molecular structure of solids and gases seems generally to be considered sharp (except for possible absorbed monolayers of molecules under certain circumstances), the interface between liquids and gases seems to have been open to other interpretations. Several texts on kinetic theory[9] have referred to a zone of variable molecular density which

---

[9] See, Boynton, reference 16, Kleeman, reference 41, and Loeb, reference 55, particularly pp. 106-113.

exists between liquids and gases. Derivation of certain thermodynamic equations from kinetic theory is based on this description.

Since so little is really known of the physical behavior of the surfaces of condensed phases in terms of molecular theory, any comments made from this viewpoint are highly speculative. In general, however, such considerations do not contradict the idea that the thickness of any transition zone between the phases has dimensions which correspond to those of a molecule itself. Fortunately, it is possible to examine this question with some indirect, but rather conclusive, experimental evidence.

According to a theory in optics developed by Fresnel, if there is an abrupt change in the refractive indices of two contacting phases, then light reflected from the phase interface at a critical angle which is a function of the refractive indices should be plane polarized. Light reflected from solid surfaces seemed to obey this law, but it had been noted that light reflected from liquid surfaces was elliptically polarized. Therefore, some doubt was expressed as to the applicability of the hypothesis made by Fresnel. This situation was examined carefully by Lord Rayleigh (74) in 1892, and he found that the ellipicity of light reflected from a water surface could be greatly reduced if precautions were taken to avoid grease films, which are almost invariably present under static conditions. Since these films are commonly accepted as monomolecular layers, Adam[10] suggests that this is almost conclusive proof that the surface of a liquid is very sharply defined, since this monolayer is responsible for almost all of the ellipicity in the reflected light. Further observations on a variety of organic liquids by Raman and Ramdas (73) agree with Rayleigh's work.

Bikerman (13) seems to suggest a gradual variation in the density between a liquid and gas phase. He claims that the sharpness of a liquid-gas interface cannot be deduced from the above evidence and refers to a paper by Bruce (19) in support of this contention. Bruce has reviewed the status of theoretical interpretations of the ellipicity of light reflected from surfaces. These are, in general, based on the assumed existence of a transition layer of thickness, $\delta$, with a refractive index which is considered constant as a first approximation. Data on the ellipicity of light reflected from a liquid surface, when interpreted in terms of these theories, do not permit calculation of $\delta$ unless the refractive index of the layer is known. If the surface layer has a refractive index which is the geometric mean of the refractive indices of the two uniform phases, then its thickness is found to be at a minimum value. For the case of pure water (evidently in air) this minimum is

[10] See reference 2, p. 5.

given as 2.26 angstrom units. If, on the other hand, the refractive index of the layer is the same as that of either uniform phase, then, according to Bruce, the transition layer must be "almost infinitely deep." It would seem in this case that the transition zone is identical with one of the phases and is therefore meaningless. If the existence of a transition layer is assumed at all, it must be considered as having a refractive index intermediate between those of the two uniform phases and, according to Bruce's analysis, as being of very small thickness approaching a molecular diameter.

The above considerations indicate that the influence of one phase on another is restricted to a sharply defined region where the two come in contact. Since the refractive index of the gas phase is uniform to the phase interface, it seems reasonable to assume that its density and other properties are also uniform. There may be a greater possibility of surface peculiarities in the liquid, since these might conceivably exist without altering the density or refractive index. That these peculiarities are restricted to the liquid phase is further supported by the observation that the surface tension appears to be a function only of the liquid and not of the gas in contact with it.

**1.4. Results of Encounters between Gas Molecules and Liquid or Solid Surfaces.** An important consideration in the theoretical treatment of interphase transfer processes is the question of what occurs when a molecule from a gas phase approaches the surface of a liquid or solid. Unfortunately, no entirely satisfactory discussion of this subject can be given at present. Detailed knowledge of the structure of solid and liquid phases, and in particular of their surface structure, is too limited to permit a purely deductive analysis. The present section will review some of the opinions which have been expressed on the problem, and the preceding section and a later discussion of the absolute rate of vaporization also bring certain evidence to light.

It has already been noted (Section 1.2) that in the absence of disturbing influences a system of gas molecules tends toward a Maxwell distribution of velocities represented by the function (1.2-9). However, even in the absence of transfer phenomena which may occur in a gas, it is apparent that the presence of a condensed phase may constitute a disturbing influence for the molecules in its neighborhood. Chapman and Cowling[11] have remarked that the rigorous derivation of the distribution function (1.2-9) is possible only for a gas enclosed in a vessel with perfectly smooth, elastic walls, so that the only effect of encounters with the surface will be a reversal of the component of molecular velocity normal to the surface.

[11] See reference 21, pp. 76-77.

# INTRODUCTION 17

As Fowler[12] suggests, however, there are at least three types of surface reflection which will preserve the Maxwell distribution of molecular velocities. The first of these is specular reflection of the kind referred to by Chapman and Cowling. The second is diffuse reflection, in which the velocity of a reflected molecule has no correlation with its velocity of incidence[13] and the probability of a molecule leaving the surface is equal for all directions. (The *rate* at which molecules having a velocity C leave the surface is proportional to the scalar product of their velocity and a unit normal to the surface, or $n \cdot C$.) This law of reflection is exactly analogous to Lambert's Cosine Law for the diffuse reflection of light. Still a third possible form of reflection which will preserve the Maxwell velocity distribution is reflection by direct reversal of path. This type, however, seems to be physically inconsistent with any reasonable description of the mechanical properties of a surface.

One of the earliest discussions of the possible forms of reflection was that of Maxwell (61) in 1879. He clearly recognized the real complexity of the problem and made it evident that his treatment was only approximate. The theory of perfectly specular reflection as a complete and sufficient explanation of molecular encounters with a surface was quickly dismissed because, according to this view, a gas can exert only a normal stress on a surface, while the behavior of real gases flowing past solids certainly contradicts this.

As an alternative, Maxwell proposed that a solid surface might be considered as consisting of hard spherical molecules so far apart that none shields any other from the impact of gas molecules. In this case, all possible directions of a gas molecule after collision with the surface are equally probable. This hypothesis is equivalent to the law of diffuse reflection mentioned above. Theoretically, the surface might be replaced with a uniform gas having the same characteristics as the gas under consideration. Maxwell also suggested that the molecules of the gas might be thought to condense and evaporate at the surface.

As a third picture of the solid surface, Maxwell proposed that its molecules may be so close together that they will to some extent shield one another from collisions with gas molecules. Then gas molecules which approach the surface obliquely will be more likely to strike surface molecules in glancing collisions. There is no longer equal probability for all directions of reflection; instead, the tangential velocity

---

[12] See reference 26, pp. 697-699.
[13] The mean properties of the incident and reflected streams must be the same, however, if the system is in equilibrium.

components of an incident molecule will tend to remain the same. It was suggested that, in effect, the condition of the molecules after colliding with the surface would be intermediate between that produced by specular and that produced by diffuse reflection. Mathematically, Maxwell treated the situation by supposing that a fraction $\nu$ of the surface reflected gas molecules specularly, while the remaining part, $1 - \nu$, reflected them diffusely. It is evident that this picture was intended to be only an approximate representation of the real situation.

As Fowler has indicated,[14] it might be desirable to correlate $\nu$ with the direction (and perhaps the speed) of incidence, but he does not pursue this idea and remarks that a treatment similar to Maxwell's is "probably adequate." Applied to an individual molecule, $\nu$ has only a statistical meaning. Maxwell's mathematical treatment requires that it be the same for all incident molecules, but his discussion of molecular processes at solid surfaces clearly recognizes the possibility that it may not be. It is important to recognize that, unless $\nu$ is independent of the velocity of incidence, a Maxwell velocity distribution near the surface is impossible. For example, consider that all the molecules which approach the surface at less than some critical angle are specularly reflected. Then the velocity components of these molecules parallel to the surface will be unchanged. On the other hand, if all molecules whose angle of approach is greater than the critical angle are diffusely reflected, then some of these molecules will after reflection have velocities which are the same as those of the specularly reflected group. The net effect of this will tend to alter the velocity distribution of any incident Maxwellian stream.

The discussion thus far has assumed that the condensed phase is nonvolatile and that gas molecules have no tendency to become permanently attached to it. Furthermore, the circumstance that the condensed phase may not be in thermodynamic equilibrium with the gas has not been considered. When this condition prevails, new questions arise. Clearly, gas molecules tend to approach equilibrium with any surface they encounter; otherwise, there could be no interphase transfer processes. The question which must be examined is whether the molecules returning to a gas from a phase interface have the same characteristics as they would have in a gas phase in complete equilibrium with the condensed phase, or, if not, to what extent they depart from such a condition.

No allowance has been made up to this point for the possibility of condensation and evaporation. In 1916 Langmuir (49) examined the question of the interaction of gas molecules with a condensed phase in

---

[14] See reference 26, p. 699.

detail from this viewpoint. He had already (48) discussed concentration discontinuities at a surface for a rather complicated combined case of diffusion and chemical reaction. On the basis of various theoretical and experimental considerations, he proposed that all gas molecules which strike the surface of a solid or liquid condense upon it and subsequently evaporate. There is an important distinction between this process and one of diffuse reflection. If the rate at which molecules return to the gas from the surface tends to correspond to the rate of return to a gas in equilibrium with the surface, then there is obviously a possibility of net condensation if the condition of the surface is such that the rate of return is slower than the rate of incidence. On the other hand, it is impossible for molecules to return to the gas faster than they condense, regardless of the condition of the surface, unless some auxiliary source of molecules is available. In this case, heat or momentum transfer may occur at the phase interface, but not mass transfer. If the surface itself consists of the same molecular specie as the gas, then it is possible for the rate of emission to exceed the rate of condensation. Net vaporization is then taking place.

The strongest support for Langmuir's theory is found in experimental evidence on the absolute rate of vaporization, which will be discussed in Chapter II. A simple experiment which Langmuir used as a bit of evidence might be mentioned here. If a sheet of mica is placed near the filament of a vacuum tungsten incandescent lamp and the bulb is allowed to discolor because of evaporation of tungsten from the filament and subsequent condensation on the glass, it is found that a sharply defined clear "shadow" of the mica is cast on the bulb. Since the molecular density in the bulb is reduced sufficiently for the mean free path of tungsten molecules to be considerably greater than the dimensions of the bulb, it would be expected that this result would be obtained, provided all or nearly all of the tungsten molecules striking the glass condense immediately and are not reflected. Langmuir recognized that this was a very special case so far as substances and temperatures were concerned. He did not mention, however, that the geometry is also exceptional, since all the molecules emanating from the center of a spherical bulb have velocities which are almost normal to the glass surface they strike. According to Maxwell's discussion, this is the circumstance where specular reflection is least expected. It would apparently have no effect, however, on the distinction between condensation and diffuse reflection.

Many elaborate theoretical analyses have been made of the encounters between gas molecules and solid surfaces. These range from the discussion of Baule (10) in 1916 in terms of classical mechanics to the more recent work of such investigators as Lennard-Jones and

Devonshire (52), Herzfeld (32), and others in terms of wave-mechanical theories. A recent paper by Wyllie (101) applies some of the modern ideas of liquid structure to a consideration of the liquid surface. These latter types of treatment seem to be the only adequate theoretical approach to the problem. The state of development of these theories at present, however, indicates that their results are obtained more conveniently and more accurately by experiment whenever possible.

**1.5. The Theory of Viscous Slip and Temperature Jump.** The transport properties of a nonuniform gas, such as its viscosity and thermal conductivity, were defined long before they could be deduced from more fundamental ideas on the structure of the gas. Similarly, there have been defined two properties associated with the discontinuities which exist in the temperature and velocity distributions of nonuniform gases near surfaces. These are the coefficients of viscous slip and temperature jump, $\xi$ and $\zeta$, which are defined by the equations

$$v_0 - v_s = \xi \frac{dv}{dx}, \qquad (1.5\text{-}1)$$

$$T_0 - T_s = \zeta \frac{dT}{dx}, \qquad (1.5\text{-}2)$$

where $v$ and $T$ denote velocity and temperature, respectively. The subscript $s$ refers to the surface with which the gas is in contact and the subscript 0 refers to the apparent condition of the gas at the surface. The derivatives are the velocity and temperature gradients in the gas some distance from the surface.

The significance of these definitions will be clearer if figures 1 and 2 are examined. The solid curves indicate the actual conditions existing in the gas in a direction normal to the surface. If heat or momentum is transported through the gas at a uniform and constant rate, then the theory of these processes as they occur far from the phase boundary requires that the velocity or temperature gradient be constant. An extrapolation based on the constant slope of these curves gives $v_0$ and $T_0$, the apparent velocity and temperature of the gas at the surface. It is evident that if the condition of the gas at some point 1 is known, this point being a distance $\Delta x$ from the surface, then the rate of heat transfer must be based on a gradient $dT/dx = (T_1 - T_0)\Delta x$, or the rate of momentum transfer must be based on a gradient $dv/dx = (v_1 - v_0)/\Delta x$. One might also base these gradients on the temperature difference $T_1 - T_s$, or the velocity difference $v_1 - v_s$, provided the distance $\zeta$ or $\xi$ is added to $\Delta x$ in each case. Because of this fact, the

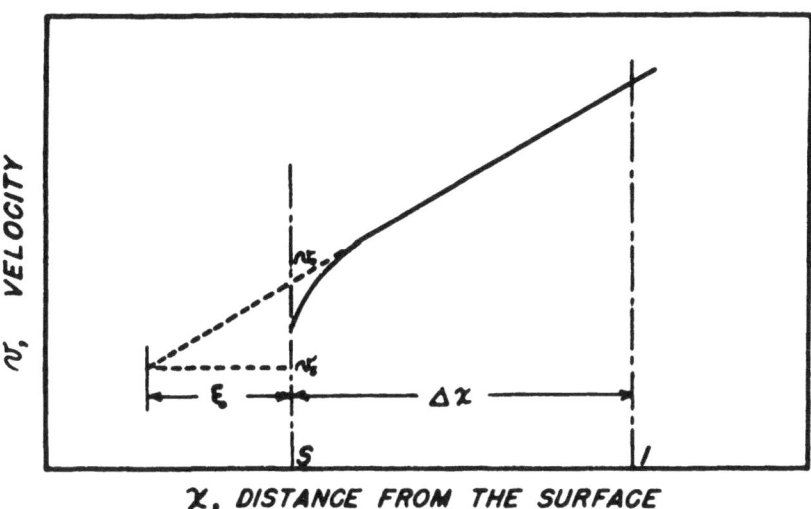

FIGURE 1.

The velocity distribution in a gas moving past a surface

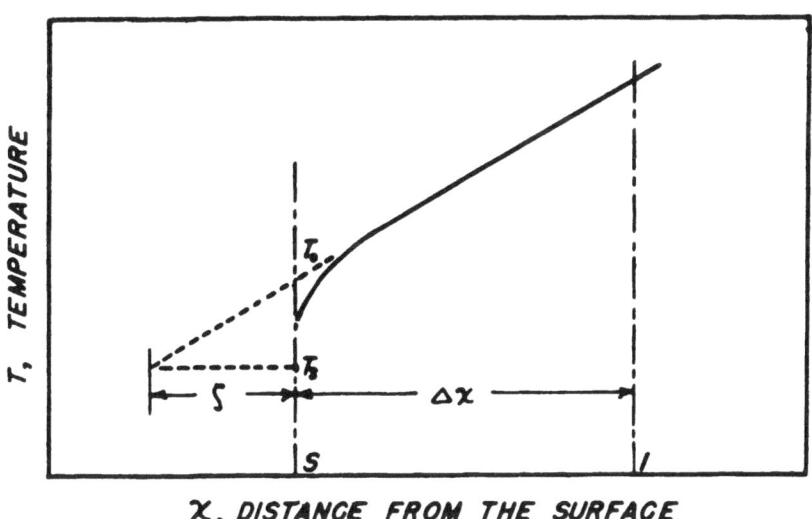

FIGURE 2.

The temperature distribution in a gas exchanging heat with a surface coefficients $\xi$ and $\zeta$ are sometimes called the viscous slip and temperature jump distances.

In 1879 Maxwell explained viscous slip in terms of a theory which evidently has not been improved.[15] First, it is assumed that the rate at which gas molecules strike a surface is the same as the rate at which they cross from one side to the other of a plane moving with a uniform gas. If we take the convention that the $x$ axis is normal to the surface and extends into the gas, this rate of flow to the surface will be denoted by $w_{0-}$ in units of mass per unit time per unit area. It is then presumed that this stream of incident gas molecules carries a total tangential component of momentum which may be divided into two parts. The first of these is $-\tfrac{1}{2}\mu dv/dx$ or one half the rate of momentum transport in the gas at some point such as 1 in figure 1. This is because momentum is being transported equally by the $w_{1-}$ and $w_{1+}$ streams at this point. The second part of the tangential momentum which $w_{0-}$ carries is due to the slip velocity of the gas past the surface, that is, $v_0 - v_s$.

A fraction $\nu$ of the total tangential momentum of $w_{0-}$ is now considered to be given to the surface on impact. This is in accordance with Maxwell's idea of combined specular and diffuse reflection, as discussed in Section 1.4. Therefore, equating the momentum surrendered to the surface and the total rate of momentum transport at 1, we obtain

$$\nu[-\tfrac{1}{2}\mu(dv/dx) + w_{0-}(v_0 - v_s)] = -\mu dv/dx. \qquad (1.5\text{-}3)$$

Substituting for $v_0 - v_s$ its value from the definition of the slip coefficient in (1.5-1) and solving for the latter, we obtain

$$\xi = -\frac{1}{2}\frac{2-\nu}{\nu}\frac{\mu}{w_{0-}}, \qquad (1.5\text{-}4)$$

and, if $w_{0-}$ and $\mu$ are expressed in terms of the mean free path, $L$, based upon the approximate condition of the gas near the surface, then (1.5-4) becomes

$$\xi = 2c\frac{2-\nu}{\nu}L, \qquad (1.5\text{-}5)$$

where $2c$ is a constant which is nearly unity. Millikan (63) has evaluated $\nu$ from available data on viscous slip and found it to vary from 0.79 to 1.0 for different surfaces (including both liquids and solids)

[15] See reference 39, p. 295.

and gases. Since $\nu$ cannot reasonably be expected to exceed unity,[16] these results tend to support the adequacy of the assumptions and approximations made in the derivation of (1.5-5). It cannot be said, however, whether the differences between $\nu$ and unity are due to an inaccurate theory of the behavior of the gas or whether $\nu$ really is less than unity. The fact that these differences amount to only 20 percent at most and always show $\nu \leq 1$ is certainly encouraging.

The phenomenon of temperature jump can be treated in a similar manner.[17] It is necessary here, however, to introduce Knudsen's (44) definition of the thermal accommodation coefficient $a$ in place of $\nu$. This may be defined in terms of either the mean kinetic energies of various classes of molecules or their associated temperatures. Choosing the latter method, we have

$$a = \frac{T_i - T_r}{T_i - T_s}, \tag{1.5-6}$$

where $T_i$ and $T_r$ are the mean temperatures of the incident and returning streams of molecules and $T_s$ is the temperature of the surface. The accommodation coefficient is often defined as a limiting value when temperature nonuniformities are small, so that any possible temperature dependency of $a$ is allowed for. An analysis very similar to the one made for viscous slip results in the following equation for the temperature jump coefficient:

$$\zeta = \frac{2-a}{a}\left(\frac{k}{\mu c_v}\right)\frac{4c}{\kappa+1}L, \tag{1.5-7}$$

where $c_v$ is the specific heat at constant volume and $\kappa$ is the ratio of the specific heats at constant pressure and constant volume.

It should be mentioned that the derivation of (1.5-7) in some respects requires more drastic assumptions than did the derivation of (1.5-5) for the slip coefficient. The energy of complex molecules is due to various kinds of motion (translation, rotation, vibration) and is not necessarily distributed in a nonuniform gas in the same way it is in a uniform gas. There is definite reason to suppose (24, 93) that the rates of energy transfer by collisions with other molecules or a surface are not equal for all types of energy. The differences are such, however, that it is probably adequate to use an average accommodation coefficient at the present level of theory on the subject.

---

[16] Unless reflection of molecules by direct reversal of path is acceptable.
[17] See reference 39, p. 312.

Values of $\nu$ and $a$ may also be obtained from experiments on free molecule flow and heat conduction. Although the concepts of viscous slip and temperature jump are not involved in such work, the quantities $\nu$ and $a$ are. The evidence obtained in these experiments substantiates the other data on the coefficients. Most of the data on thermal accommodation coefficients are based on experiments of this sort. These data are summarized in part by Kennard,[18] and other data are given by Wiedman and Trumpler (99). The thermal accommodation coefficient $a$, like $\nu$, has always been found to be less than or equal to unity. In some cases, however, it has varied down to considerably lower values and also shows certain anomalies in its behavior, depending upon how the experimental work is conducted. Such data as are available are chiefly for very volatile substances (such as hydrogen or helium) in contact with metallic solids. These cases have little application to the present study. Limited experimental data for vapors and surfaces of the same molecular specie seem to indicate that for these cases the accommodation coefficient is close to unity.

[18] See reference 39, p. 323.

## CHAPTER II

## THE ABSOLUTE RATE OF VAPORIZATION OF A PURE SUBSTANCE

**2.1. Systems at Equilibrium.** Since the earliest days of kinetic theory, the equilibrium between a liquid or solid and its vapor has been visualized not as a static state, but rather as a process of vaporization and condensation occurring at equal rates. The first to examine this idea quantitatively was Hertz (31) in 1882. His pioneer paper presented an equation which has been used almost universally to calculate the absolute rate of vaporization of pure liquids and solids. Important comments on the assumptions made in deriving it and on the extent of its applicability to actual experimental work have been added over subsequent years. These will be reviewed here with certain refinements and additions.

It will be assumed in the following discussion that, for equilibrium cases,
(a) any uncertainty in the precise location of the phase interface is small with respect to a mean free path in the gas phase, and
(b) above this interface there exists a uniform gas with the velocity and internal energy of its molecules distributed according to the same laws which are known to apply to a uniform gas in the absence of disturbing influences.

The fact that the interface may be considered in the manner indicated was discussed in Section 1.3. The characteristics for encounters between gas molecules and the phase interface necessary so that the second assumption is met were discussed in Section 1.4. The propriety of this representation of the gas in the neighborhood of the phase interface will be left to be resolved chiefly by the agreement between the theory based on it and experiment.

For thermodynamic equilibrium to exist between two phases, it is a necessary condition that their temperatures be the same, since otherwise there would be a flow of heat between them. It is possible, however, for a condensed phase to exist under a pressure other than its own vapor pressure and to have a gas phase in equilibrium with it. Such a condition may be ideally imagined through the application of pressure to the surface of the condensed phase by a membrane which is permeable to single (gas) molecules, but not to aggregates of (liquid or solid) molecules. It might also be imagined in terms of a hypothetical gas which is present with the vapor, but is completely insoluble in the liquid or solid and does not affect the behavior of the vapor in any way. This gas may exert a pressure on the surface in excess of the pressure of the vapor which is mixed with it.

A thermodynamic relation between the vapor pressure of a liquid and the actual pressure under which it exists was first derived by Thomson (88) for a certain special application. A more satisfactory derivation for the present purpose is given by Keenan.[1] The result of this analysis shows that if the vapor behaves as a perfect gas and the liquid is incompressible, then the vapor pressure, $\bar{P}_s$, of a liquid under pressure $P_s$ is related to the normal vapor pressure, $P_s^*$ (which is a function of temperature only), by

$$\bar{P}_s = P_s^* \exp\left(\frac{P_s - P_s^*}{\gamma_s R T_s}\right), \qquad (2.1\text{-}1)$$

where $\gamma_s$ and $T_s$ are the density and temperature of the liquid phase and $R$ is the perfect gas constant. This equation might also be written in terms of the equivalent mass densities as

$$\tilde{\gamma}_s = \gamma_s^* \exp\left(\frac{P_s - P_s^*}{\gamma_s R T_s}\right). \qquad (2.1\text{-}2)$$

It is apparent from these equations that, unless the pressure on a liquid surface is greatly different from its vapor pressure, the effect of pressure on the vapor pressure is very small.

Throughout this book the tilde is used to indicate the properties of a gas phase which would be in equilibrium with the surface of a solid or liquid phase as it exists. The asterisk is used to denote the properties of a gas phase in equilibrium with a liquid or solid under its own vapor pressure. The distinction between the two is usually slight and has not been noted in previous treatments of interphase mass transfer.

On the basis of assumption (b) at the beginning of this section, we may proceed to calculate the rate at which mass leaves the phase interface and enters the gas. The convention to be used throughout this book is that the origin of the coordinate axes is at the phase interface and the $x$ axis is directed into the gas. Denoting the mass rate of flow per unit area from the phase interface by $w_{s+}$, we have from (1.2-4)

$$w_{s+} = \int_{-\infty}^{\infty}\int_{-\infty}^{\infty}\int_{0}^{\infty} m U \tilde{f}_s \, dU dV dW, \qquad (2.1\text{-}3)$$

[1] See reference 38, p. 471.

where $\bar{f}_s$ is the distribution function for the gas in equilibrium with the surface. Since the gas has been considered uniform, $\bar{f}_s$ is given by (1.2-9a) with $\bar{U} = \bar{V} = \bar{W} = 0$ because of the equilibrium conditions. Then (2.1-3) becomes

$$w_{s+} = \tilde{\gamma}_s \left(\frac{\beta_s^3}{\pi^{3/2}}\right) \int_{-\infty}^{\infty} \int_{-\infty}^{\infty} \int_0^{\infty} U e^{-\beta_s^2 [U^2 + V^2 + W^2]} dU dV dW, \quad (2.1\text{-}4)$$

which is easily evaluated[2] and found to be

$$w_{s+} = \tilde{\gamma}_s \left(\frac{\beta_s^3}{\pi^{3/2}}\right)\left(\frac{1}{2\beta_s^2}\right)\left(\frac{\pi^{1/2}}{\beta_s}\right)\left(\frac{\pi^{1/2}}{\beta_s}\right) \quad (2.1\text{-}5)$$

$$= \frac{\tilde{\gamma}_s}{2\pi^{1/2} \beta_s}. \quad (2.1\text{-}6a)$$

From the value of $\beta$ given by (1.2-8), this may be alternatively written

$$w_{s+} = \tilde{\gamma}_s \sqrt{\frac{RT_s}{2\pi M}} \quad (2.1\text{-}6b)$$

$$= \tilde{P}_s \sqrt{\frac{M}{2\pi RT_s}}. \quad (2.1\text{-}6c)$$

Let us now consider that of those molecules streaming from the liquid or solid surface a fraction $\sigma$ originate because of spontaneous processes inherent in the surface. (For any given state of the surface there will be a certain number of molecules having characteristics which will enable them to leave and enter the gas. Provided the state is maintained, this number and the rate at which they leave the surface will be constant.) This spontaneous part of the stream of molecules from the surface, $\sigma w_{s+}$, is known as the absolute rate of vaporization. The remaining part, $(1 - \sigma)w_{s+}$, is due to a "reflection" of gas molecules which do not enter the liquid or solid phase.

The quantity $\sigma$ is known variously in the literature as a vaporization,[3] evaporation, condensation, or accommodation coefficient. The last of these terms is inappropriate and will not be used here. The name "condensation" coefficient arises because $-\sigma w_{s+} = \sigma w_{s-}$ is

[2] Appendix A lists integrals which occur frequently in this study.
[3] This should not be confused with the empirical Stefan-Mache vaporization coefficient defined in (1.1-1).

obviously the absolute rate of condensation, and $\sigma$ will be identified by this name in the following pages. As mentioned in Section 1.4, it is necessary that $\sigma$ be constant and independent of the vector velocity of incidence or emission of gas molecules at the surface, for only then can the Maxwell velocity distribution reasonably be expected to be preserved. Until demonstrated otherwise, however, the condensation coefficient may be regarded as having a functional dependence upon the state of the surface and the kind of molecules involved.

**2.2. Systems at Nonequilibrium.** All of the considerations of the preceding section apply to a system which is in equilibrium, so that neither energy nor mass cross the phase interface at a net rate. Let us now examine the case when the gas phase has nonequilibrium characteristics. There is no reason to suppose that the state of the gas has any influence on the *absolute* rate of vaporization at the surface, which is reasonably expected to be a function only of the state of the liquid or solid surface.

The collisions of gas molecules with the surface determine the pressure there (see Section 1.2d), so that their effect from this viewpoint is merely one of the factors which define the surface state. Of course, the distribution of momentum for incident and leaving molecules as well as their mean value of momentum might conceivably affect the absolute rate of vaporization, but it hardly seems likely that this could be of any quantitative significance. Even the magnitude of the normal surface pressure has a very small effect on the vapor pressure of liquids, according to (2.1-1). The energy of molecules in the gas phase is also irrelevant unless this energy is communicated to the liquid or solid molecules, in which case it again becomes associated with the state of the surface.

There is one consideration which will have some effect on the absolute rate of vaporization. This is the fact that if interphase mass transfer is occurring, then the mean velocity of liquid or solid molecules at the surface is not zero with respect to the coordinate axes which have been located there. This will probably cause the absolute rate of vaporization as well as the velocity distribution of molecules emitted from the surface to be otherwise than they are in an equilibrium case. However, the density of the liquid or solid phase is ordinarily so great compared with that of the gas (especially when the theories of this study have a practical importance) that neglect of this mean velocity introduces no large error. Therefore, the effect of the mean velocity of liquid or solid molecules at the phase interface is neglected.

The absolute rate of vaporization is thus considered to be a function only of the thermodynamic properties (for example, temperature and pressure) of the liquid or solid surface and is $\sigma w_{s+}$, where $w_{s+}$ is given by (2.1-6a), (2.1-6b), or (2.1-6c). Furthermore, the distribution of velocities for the molecules emitted because of absolute vaporization is given by $\sigma \bar{f}_s$, where $f_s$ is the distribution function for the vapor which would be in equilibrium with the surface.

It is of interest to remark here that Penner (68, 69) has derived an equation for the absolute rate of vaporization of a liquid from the theory of absolute reaction rates. This derivation results in an expression which is identical in form with that given in the preceding section. Melville (62) has shown theoretically that the absolute rate of vaporization of a solid is unaffected by small irregularities in its surface and can be based on the apparent surface area, provided the condensation coefficient is unity. If $\sigma < 1$, then a correction, which is generally small, is needed.

**2.3. Experimental Evidence.** There have been several experimental investigations of the absolute rate of vaporization of a pure substance. The following requirements are essential for valid determinations, although, unfortunately, they have often been neglected:
(a) an accurate estimate of the surface temperature of the evaporating substance,[4]
(b) a knowledge of the interfacial area,
(c) a surface which will condense all the evaporating molecules without reflection, and
(d) a distance between the vaporizing and condensing surfaces which is small compared with the gas mean free path.

The necessity for the first two requirements is immediately apparent. Requirements (c) and (d) are needed to prevent the return of any evaporating molecules to the surface of the pure substance. In order to meet requirement (d), it is necessary to reduce the density of molecules in the space between the vaporizing and condensing surfaces as much as possible. When this is done, the mean free path of gas molecules is made large and the condenser can be located far enough from the evaporating surface that there is no danger of direct contact. Also, as the experiment progresses, the changing distance between the two surfaces must not become so great that it violates requirement (d).

These essential features for valid experimental determinations impose severe practical restrictions on the number of substances suitable for examination. They unfortunately preclude the use of volatile materials, which are of great practical importance. The absolute rate of

[4] The effect of surface pressure has always been neglected, so that $\bar{P}_s$ is taken as equal to $P_s{}^*$. This assumption is justifiable in this work because all the pressure terms are small and the difference between them is negligible.

vaporization of these substances is so high under ordinary conditions that it is impossible to supply mass and energy to the surface at the required rates without violating all the necessary conditions for experimental work. Investigations have therefore been limited to such nonvolatile substances as metals at various temperatures and other materials at very low temperatures.

The first experimental study, as has already been remarked, was that of Hertz (31) in 1882. He examined the absolute rate of vaporization of mercury at ordinary temperatures and found that, when his data were interpreted in terms of an equation similar to (2.2-6c), the value of the condensation coefficient, $\sigma$, was about 1/9. Marcelin (59, 60) in 1914, evidently unaware of Hertz's work, performed some similar experiments on a variety of substances and obtained approximately the same value for $\sigma$. These early experimental arrangements were not, however, suited for accurate measurements.

In 1915 Knudsen (45) investigated the absolute rate of vaporization of mercury with great care. At first he obtained a value of $\sigma = 0.0005$, but, upon examining the surface of the mercury, he noted a slight brownish scum which he suspected was interfering with the vaporization process. Therefore, he repeated his experiments with a purer sample of mercury and obtained a value for $\sigma$ which was close to that of Hertz. Encouraged by this improvement, he used the purest mercury which he could prepare and found $\sigma = 1$ with a stated experimental error of 1 percent. This work was done much more carefully than the earlier experiments of Hertz and is consequently much more reliable.

Bennewitz (11) in 1919 studied the absolute rate of vaporization of cadmium and obtained results similar to those of Knudsen. Although he was able to increase $\sigma$ very much from the low values which were initially observed, even with the purest cadmium he was not able to obtain values of $\sigma$ greater than approximately 0.7.

Another very careful investigation of the absolute rate of vaporization of mercury was made by Volmer and Estermann (95) in 1921. They obtained values of $\sigma$ by two somewhat different experimental arrangements. Furthermore, they studied its variation with temperature for both the liquid and the solid state. The conclusions drawn from this work were that $\sigma$ for liquid mercury is equal to unity and independent of the temperature, while for solid mercury it is somewhat less than unity and varies slightly with temperature.

In 1946 Tschudin (90) reported an interesting series of observations on the absolute rate of vaporization of ice at temperature from −60 to −85°C. This seems to be the only reliable information available on a nonmetallic substance. The value of $\sigma$ is reported as $0.94 \pm 0.06$, with no variation due to temperature. The description of the experimental procedure indicates that this work is as reliable as that of Knudsen, Bennewitz, and Volmer and Estermann.

Langmuir (37, 47, 49, 50) has made extensive use of the theory of the absolute rate of vaporization to calculate the vapor pressure of such nonvolatile metals as tungsten, molybdenum, platinum, nickel, iron, copper, and silver. In this work, the value of $\sigma$ was assumed to be unity, and from the rate of vaporization the vapor pressure was calculated. In general, the results of this work agree approximately with vapor pressure data obtained by other methods.

It is significant to note that the values of $\sigma$ which can be accepted as most reliable are all close to unity. The lower values which have been occasionally reported all seem to have been obtained in experiments involving questionable techniques.[5] It cannot conclusively be said, however, on the basis of data currently available, whether the condensation coefficient is always near unity.

[5] The experimental work of Wyllie (101) might be cited as a recent example of such a case. In these experiments a small quantity of glycerine was contained in a tube capped with a metal plate containing a small hole. This tube was then placed in a vessel connected to a liquid air trap and a vacuum system. The rate of effusion of glycerine from the hole was then used to calculate its vapor pressure, according to a procedure originally suggested by Knudsen (43).

The cap was then removed from the tube and evaporation allowed to proceed from the free surface of the glycerine. $\sigma$ was calculated from the equation

$$\sigma = \frac{\text{rate of evaporation from free surface}}{\text{rate of evaporation (effusion) from hole}} \times \frac{\text{area of hole}}{\text{area of free surface}}.$$

This equation is correct only if there is no condensation at the free surface of the glycerine. However, the design of the apparatus makes it apparent that a considerable number of the evaporating molecules will be returned to the liquid, owing to collisions with the apparatus, if not with themselves, before they reach the liquid air trap. Thus, it is not surprising that a low value ($\sigma$ = 0.052) is reported for the "condensation" coefficient.

# CHAPTER III

## THE SIMPLE THEORY OF INTERPHASE MASS TRANSFER OF A PURE SUBSTANCE

**3.1. The Absolute Rate of Condensation under Nonequilibrium Conditions.** If evaporation or condensation is occurring, then there must be a mass motion of gas from or to the phase interface. Provided no intraphase transfer process is taking place in the gas, then some distance from the interface it should be moving as a uniform gas. Even if some process such as heat transfer is occurring, the deviation from the Maxwell velocity distribution will ordinarily be small. In this chapter it will be assumed as a first approximation that the velocity distribution of a uniform gas in simple mass motion prevails very near to the interface of the two phases.

On this basis, the mass rate of flow of those molecules which are moving toward the phase interface ($U < 0$) may be computed from (1.2-4) and is

$$w_{0-} = \int_{-\infty}^{\infty} \int_{-\infty}^{\infty} \int_{-\infty}^{0} mUf_0 \, dU \, dV \, dW, \qquad (3.1\text{-}1)$$

where the subscript 0 denotes the properties of the alleged uniform gas at the interface. Taking the distribution function as (1.2-9a) with $\bar{U} = U_0$, corresponding to the rate of mass transfer, and $\bar{V} = \bar{W} = 0$, we have

$$w_{0-} = \gamma_0 \int_{-\infty}^{\infty} \int_{-\infty}^{\infty} \int_{-\infty}^{0} U \left(\frac{\beta_0^3}{\pi^{1/2}}\right) e^{-\beta_0^2[(U-U_0)^2 + V^2 + W^2]} dU \, dV \, dW. \qquad (3.1\text{-}2)$$

Integrating with respect to $V$ and $W$, we obtain

$$w_{0-} = \frac{\gamma_0 \beta_0}{\pi^{1/2}} \int_{-\infty}^{0} U e^{-\beta_0^2 (U - U_0)^2} dU. \qquad (3.1\text{-}3)$$

The integral in (3.1-3) may be evaluated as follows:

$$\int_{-\infty}^{0} U e^{-\beta_0^2 (U - U_0)^2} dU = \int_{-\infty}^{-U_0} U e^{-\beta_0^2 (U - U_0)^2} d(U - U_0) \qquad (3.1\text{-}4)$$

$$= \int_{-\infty}^{-U_0} (U - U_0) e^{-\beta_0^2 (U - U_0)^2} d(U - U_0)$$

$$+ U_0 \int_{-\infty}^{-U_0} e^{-\beta_0^2 (U - U_0)^2} d(U - U_0) \quad (3.1\text{-}5)$$

$$= \frac{1}{2} \int_{\infty}^{U_0^2} e^{-\beta_0^2 (U - U_0)^2} d[(U - U_0)^2]$$

$$+ U_0 \int_{U_0}^{\infty} e^{-\beta_0^2 (U - U_0)^2} d(U - U_0) \,. \quad (3.1\text{-}6)$$

The first integral in (3.1-6) is given by equation (A7) in Appendix A, while the second may be written as two probability or error integrals similar to those in equation (A1):

$$\int_{-\infty}^{0} U e^{-\beta_0^2 (U - U_0)^2} dU = -\frac{e^{-\beta_0^2 U_0^2}}{2\beta_0^2}$$

$$+ \frac{U_0 \pi^{1/2}}{2\beta_0} \left[ \frac{2}{\pi^{1/2}} \int_0^\infty e^{-x^2} dx - \frac{2}{\pi^{1/2}} \int_0^{\beta_0 U_0} e^{-x^2} dx \right] \quad (3.1\text{-}7)$$

$$= -\left\{ \frac{e^{-\beta_0^2 U_0^2}}{2\beta_0^2} - \frac{U_0 \beta_0 \pi^{1/2}}{2\beta_0^2} [1 - \Phi(\beta_0 U_0)] \right\}, \quad (3.1\text{-}8)$$

where $\Phi$ is the probability function. Substituting (3.1-8) in (3.1-3) gives

$$w_{0-} = -\frac{\gamma_0}{2\beta_0 \pi^{1/2}} \left\{ e^{-\beta_0^2 U_0^2} - \beta_0 U_0 \pi^{1/2} [1 - \Phi(\beta_0 U_0)] \right\}. \quad (3.1\text{-}9)$$

If a correction factor, $\Gamma$, is defined as

$$\Gamma = e^{-\beta_0^2 U_0^2} - \beta_0 U_0 \pi^{1/2} [1 - \Phi(\beta_0 U_0)], \quad (3.1\text{-}10)$$

(3.1-9) becomes simply

$$w_{0-} = -\frac{\gamma_0}{2\beta_0 \pi^{1/2}}\Gamma, \qquad (3.1\text{-}11)$$

and, noting (2.1-6a), we may express (3.1-11) in terms of $w_{s+}$ as

$$w_{0-} = -\left(\frac{\gamma_0}{\tilde{\gamma}_s}\right)\left(\frac{\beta_s}{\beta_0}\right)w_{s+}\Gamma, \qquad (3.1\text{-}12)$$

or, since $\beta$ is inversely proportional to the square root of the absolute temperature (1.2-8), as

$$w_{0-} = -\left(\frac{\gamma_0}{\tilde{\gamma}_s}\right)\left(\frac{T_0}{T_s}\right)^{1/2}\Gamma w_{s+}. \qquad (3.1\text{-}13)$$

The rate at which mass strikes the phase interface is given by (3.1-13) for any condition of the gas. If the gas is in equilibrium with the surface, then $U_0 = 0$ and $\Gamma = 1$. The use of the factor $\Gamma$ was suggested in a paper by Slepian and Brubaker (83–85) in 1940. They used it in a limited context to calculate the rate of mercury condensation in connection with a theory for a special case of assumed total condensation in an ignitron tube. It has apparently not been used elsewhere in the literature to discuss the general problem of vaporization and condensation.

The correction factor $\Gamma$ has been given in (3.1-10) as a function of $\beta_0 U_0$. This latter quantity may be denoted by $\phi_0$ and conveniently expressed in terms of the other dimensionless ratios used in (3.1-13):

$$\phi_0 = \beta_0 U_0 = \frac{\gamma_0 U_0}{2\pi^{1/2}}\left(\frac{2\pi^{1/2}\beta_s}{\tilde{\gamma}_s}\right)\left(\frac{\beta_0}{\beta_s}\right)\left(\frac{\tilde{\gamma}_s}{\gamma_0}\right) \qquad (3.1\text{-}14)$$

$$= \frac{1}{2\pi^{1/2}}\frac{w}{w_{s+}}\left(\frac{T_0}{T_s}\right)^{-1/2}\left(\frac{\gamma_0}{\tilde{\gamma}_s}\right)^{-1}. \qquad (3.1\text{-}15)$$

For circumstances in which the state of the gas does not depart appreciably from its equilibrium condition, it is permissible to use the approximate equation

$$\phi_0 = \frac{1}{2\pi^{1/2}}\frac{w}{w_{s+}}. \qquad (3.1\text{-}15a)$$

The factor $\Gamma$ is calculated as a function of $\phi$ in Appendix B. Figure 3 presents the result of this calculation in graphical form for the

# THE SIMPLE THEORY OF INTERPHASE MASS TRANSFER

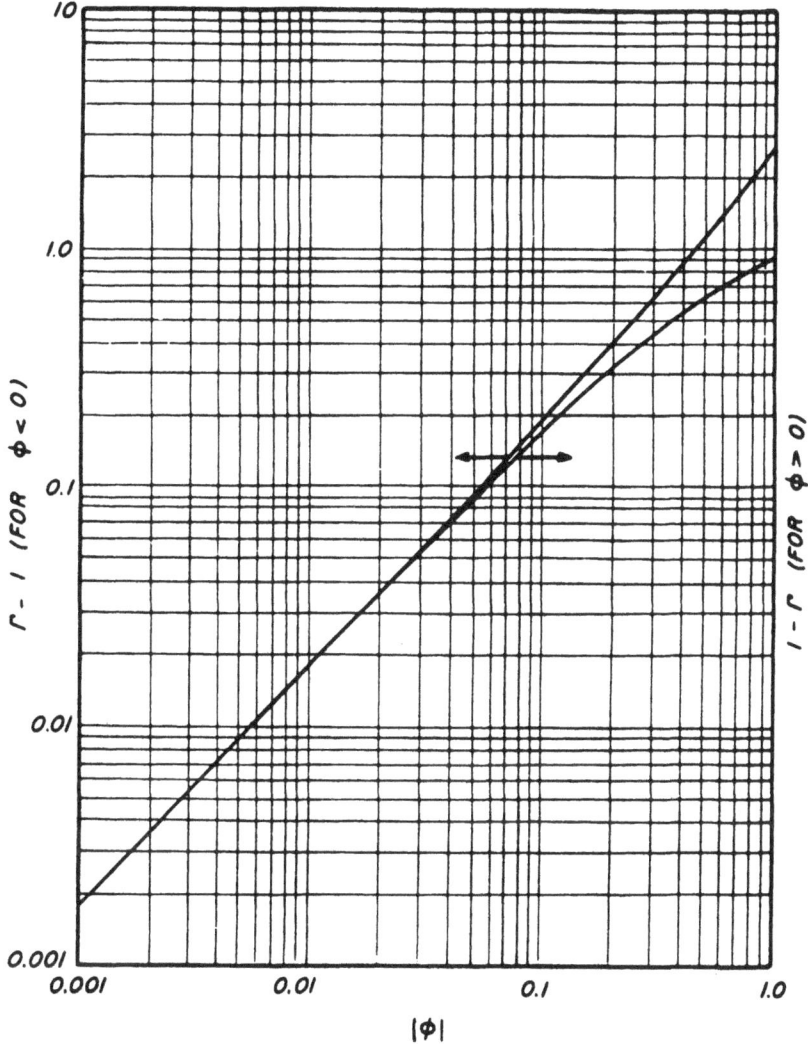

FIGURE 3.

The correction factor $\Gamma$ as a function of $\phi$

range of values $1 \geq |\phi| \geq 10^{-3}$. For smaller values of $\phi$ the approximate equation

$$\Gamma = 1 - \phi \pi^{1/2} \qquad (3.1\text{-}16)$$

gives $\Gamma - 1$ to at least four significant figures.

The absolute rate of condensation at the phase interface depends upon (3.1-13) and the fractional extent to which molecules striking the interface condense there. In this connection, the fact that the condensation coefficient has been assumed (see Section 2.1) to be independent of the vector velocity of incident gas molecules has an important consequence. Since $\sigma$ has a constant value for all gas molecules, it follows that it must be the same for all conditions of the gas. Therefore, to be consistent with Chapter II, $\sigma$ is regarded as a function only of the molecular specie under consideration and the state of the liquid or solid surface.

**3.2. Interphase Mass Transfer of a Pure Substance.** The rate of mass transfer at the phase interface is simply the sum of the absolute rates of vaporization and condensation,

$$w = \sigma w_{s+} - \sigma \left(\frac{\gamma_0}{\tilde{\gamma}_s}\right)\left(\frac{T_0}{T_s}\right)^{½} \Gamma w_{s+}, \tag{3.2-1}$$

or, written in dimensionless form,

$$\frac{w}{w_{s+}} = \sigma \left[1 - \left(\frac{\gamma_0}{\tilde{\gamma}_s}\right)\left(\frac{T_0}{T_s}\right)^{½} \Gamma \right]. \tag{3.2-2}$$

It is usually not worth while to distinguish between $\tilde{\gamma}_s$ and $\gamma_s^*$. The real pressure on a surface is in most practical instances very nearly its vapor pressure, so that this approximation is quite adequate and avoids the necessity for discussion of an approximate method for determining the real pressure on the surface. Then (3.2-2) becomes

$$\frac{w}{w_{s+}} = \sigma \left[1 - \left(\frac{\gamma_0}{\gamma_s^*}\right)\left(\frac{T_0}{T_s}\right)^{½} \Gamma \right]. \tag{3.2-3}$$

A considerable number of investigations (5, 8, 15, 18, 20, 22, 27, 33, 71, 72, 78, 89, 94, 96) have used an equation for interphase mass transfer which is usually written

$$w = \sigma \sqrt{\frac{M}{2\pi RT}} (P_s^* - P_0). \tag{3.2-4}$$

This is evidently the only equation similar to (3.2-3) which has been

THE SIMPLE THEORY OF INTERPHASE MASS TRANSFER 37

used in the literature, and it is identical to (3.2-3) only when $\Gamma$ and $T_0/T_s$ are taken as identically equal to unity.[1]

There is no reason why the temperature of the gas phase should necessarily be the same as that of the liquid or solid surface in all cases. Consider the density-temperature diagram of a typical pure substance given by figure 4. When a liquid or solid phase exists under its

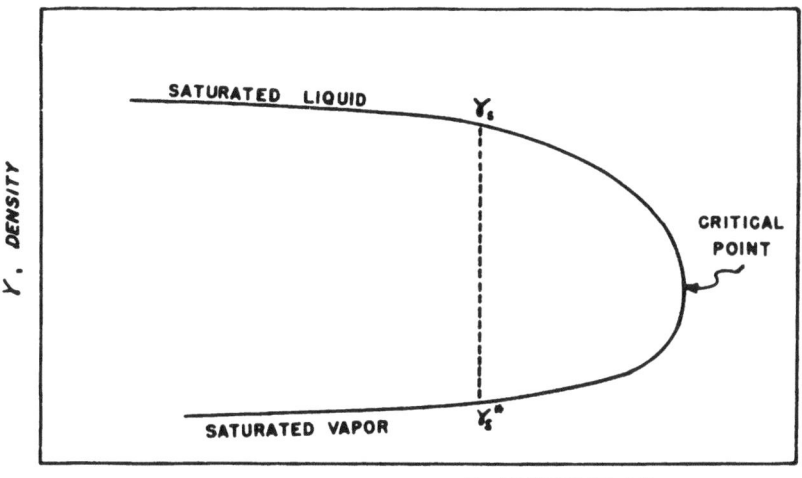

FIGURE 4.

The density-temperature diagram of a pure substance

own vapor pressure, then a condition of thermodynamic equilibrium between the condensed and gas phases will be represented by two points such as $\gamma_s$ and $\gamma_s^*$, connected by an equilibrium tie-line and lying on the saturation curves. Now, if the condition of one phase is held constant, it is clear that the other may depart from equilibrium by changing either its density or its temperature.[2] Both changes may occur simultaneously, of course, and in this case it is possible to have nonequilibrium without mass transfer. Thus, $w$ may be zero (and $\Gamma = 1$); then the

---

[1] Bosnjakovic (14) in 1932 presented an equation equivalent to (3.2-3) with $\Gamma \equiv 1$. However, as Prüger (72) has remarked, his discussion does not distinguish between interfacial and intraphase (film) temperature differences. Kirshbaum's (40) qualitative extension of some of Bosnjakovic's ideas to binary systems shows a similar confusion.

[2] A change in the density of the liquid (or solid) phase at constant temperature affects its equilibrium characteristics very slightly. This fact is related to the use of the approximate equation (3.2-3) rather than (3.2-2).

## 38 THE SIMPLE THEORY OF INTERPHASE MASS TRANSFER

temperatures and densities of the two phases must satisfy the equation

$$\left(\frac{\gamma_0}{\tilde{\gamma}_s}\right)\left(\frac{T_0}{T_s}\right)^{1/2} = 1. \tag{3.2-5}$$

The fact that the system is not in equilibrium implies an energy exchange or interphase heat transfer process.

The analysis of interphase mass transfer which led to (3.2-2) or (3.2-3) is in some respects similar to the theory of thermal transpiration between two gases, a phenomenon discovered by Reynolds (77) in 1879. Imagine two gases separated from each other by an insulating partition with a small hole less than a mean free path in diameter. If they consist of the same kind of molecules, then the rate at which each effuses through the hole will be proportional to its pressure divided by the square root of its absolute temperature. This view is based on the assumption that the velocity distribution of a stationary uniform gas exists on both sides of the partition, and this assumption is permissible as long as only small rates of mass transfer occur through the hole. If $P_I/\sqrt{T_I} = P_{II}/\sqrt{T_{II}}$, no net mass is transferred through the hole, but the energy is transpiring because of the difference between the energies of the two molecular streams. The similarity of this situation to that described by (3.2-5) is apparent.

### 3.3. Interfacial Nonequilibrium for the Special Case of $T_0/T_s = 1$.

If all the energy transfer necessary for condensation or evaporation occurs through the condensed rather than through the gas phase, then it might be expected that the interfacial temperatures of the two phases are nearly equal. This is probably the only justification which can be given for using an equation similar to (3.2-4), although restriction to this special case does not seem to have been specifically indicated in any previous work. In fact, some investigators speak of interfacial temperature differences and still use (3.2-4). It will be shown later (Chapter V) that, even for the special case of no intraphase heat transfer in the gas, on the basis of our present understanding of interphase processes this assumption of identical temperatures is probably not satisfactory except as $w/w_{s+}$ approaches zero.[3]

---

[3] There has been much inconclusive discussion on the temperature of evaporating vapors. A series of papers by Schreber (80, 81), Knoblauch and Reiher (42), Pollitzer (70), and later Möbius (66) disputed the question of the temperature of vapor evaporated from solutions of salts in water. In a later paper Schreber (82) evidently continues his side of the argument, but the question now becomes involved with considerations of intraphase heat transfer in boiling liquids, which is irrelevant to the above discussion.

Regardless of any energy transfer implications which may be suggested, it will be of interest at this point to present some detailed results for the special case of $T_0/T_s = 1$ and $\sigma = 1$. If the condensation coefficient, $\sigma$, is less than unity, then deviations from equilibrium larger than those given here are to be expected. Figures 5 and 6 give corresponding values of $y_0/\tilde{y}_s$ and $w/w_{s+}$, according to both the theory

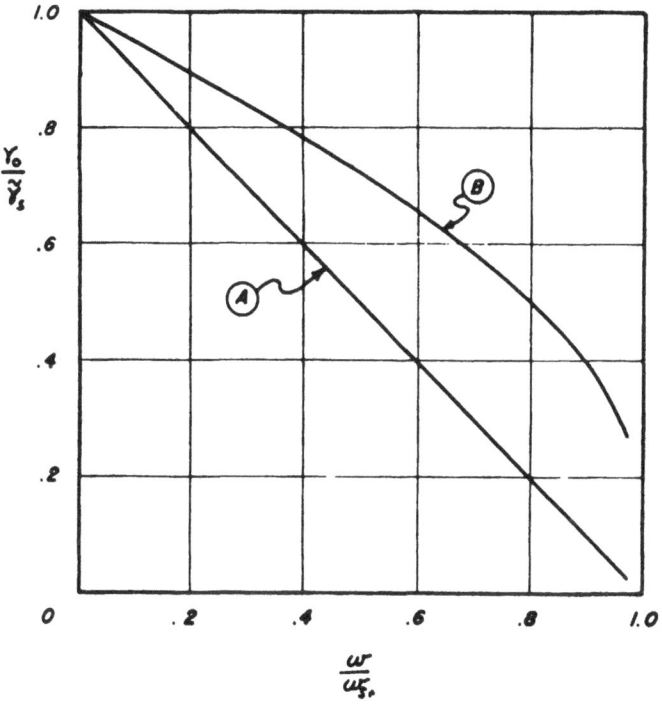

FIGURE 5.

Interfacial nonequilibrium in evaporation processes. Values of $w/w_{s+}$ and $y/\tilde{y}_s$ for the special case of $T_0/T_s = 1$ and $\sigma = 1$, calculated according to the theory of Chapter III. Curve A: equation (3.2-2) with $\Gamma = 1$; curve B: equation (3.2-2)

---

Kennard (reference 39, p. 79), on the basis of the Boltzmann distribution law (which applies to equilibrium systems), claims that "freshly formed vapor . . . ought to have the temperature of the surface of the evaporating liquid or solid." It seems obvious that an evaporating substance is not a system in equilibrium and that any application of the Boltzmann law to it is inappropriate.

Little (54) proposed to show that, even when a liquid and vapor in equilibrium, "the vapor temperature may be two thirds of the liquid temperature." For some reason which is not entirely clear, "a thermometer will never detect this temperature difference."

## 40 THE SIMPLE THEORY OF INTERPHASE MASS TRANSFER

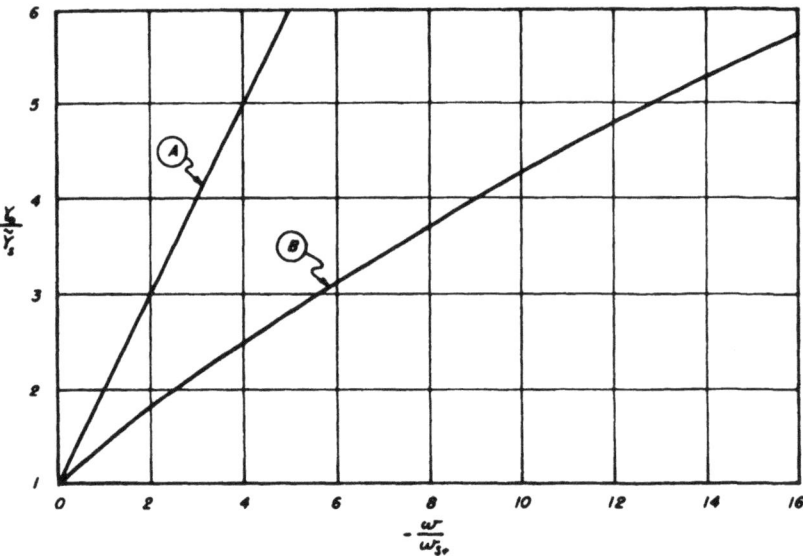

FIGURE 6.

Interfacial nonequilibrium in condensation processes. Values of $-w/w_{s+}$ and $\gamma_0/\bar{\gamma}_s$ for the special case of $T_0/T_s = 1$ and $\sigma = 1$, calculated according to the theory of Chapter III. Curve A: equation (3.2-2) with $\Gamma = 1$; curve B: equation (3.2-2)

equivalent to (3.2-4) ($\Gamma = 1$) and the theory which led to (3.2-2). Detailed calculations are given in Appendix C.

It will be noted that the newer theory incorporating the factor $\Gamma$ predicts smaller deviations from equilibrium for the same values of $w/w_{s+}$ than does the older theory. As shown in Appendix C, as $w/w_{s+} \to 0$, departures from equilibrium according to the newer theory are just one half what they would be with $\Gamma$ identically equal to unity. Neither of these theories can be expected to give accurate values of $\gamma_0/\bar{\gamma}_s$ as $w/w_{s+} \to 1$, corresponding to the absolute rate of vaporization. The limiting value of $\gamma_0/\bar{\gamma}_s$ must be 0.5 for this case. It will be shown rigorously in Section 5.7 that there is a critical value of $w/w_{s+}$ above which the gas phase does not resemble an ordinary gas. In effect, this is the limiting value for interphase mass transfer. Higher values of $w/w_{s+}$ correspond to molecular evaporation. The specific value of $w/w_{s+}$ at this critical point depends upon the exact description of the gas at the phase interface; and, since the theories of Chapters III and IV are not based on exact descriptions, they define no critical point.

**3.4. Experimental Studies of Interphase Mass Transfer.** The only experimental work which will be reviewed here is that which has been undertaken specifically as a study of interphase mass transfer and in which intraphase processes were of only incidental importance. With the exception of the work cited in this section, experimental investigations of "interphase" mass transfer have in general been considerations of intraphase transfer processes, interpreted on the basis of assumed thermodynamic equilibrium at the phase interface. Several studies on the evaporation of small droplets are exceptions to this statement. The interpretation of these is based on a theoretical paper by Fuchs (27) which involves use of (3.2-4). This work embodies so many considerations extraneous to the subject of this study, however, that it will not be discussed.

The several investigations of Alty and coworkers (3–9) and Prüger (71) were made with the specific purpose of interpretation in terms of theories of true interphase mass transfer. Alty's work was done at low pressures and Prüger's at atmospheric pressure. The latter was a student of Mache (see Section 1.1), and, since his approach to the problem was a refinement of Mache's method, this earlier work need not be examined here.

According to his first paper on the subject (3) in 1931, Alty evidently wished to make a direct determination of the absolute rate of vaporization of water and to do this by experiments similar to those of Knudsen on mercury (see Section 2.3). As has already been stated, this is experimentally impractical, a fact which soon became evident. Alty then decided to perform his experiments at pressures high enough to reduce the rate of vaporization below its maximum value and then extrapolate the results to zero pressure. A value of $\sigma = 0.01$ in (2.1-6c) was obtained as a result of this procedure.

These extrapolations are so uncertain, however, that even if the measured quantities are pertinent to the rate of vaporization (which is doubtful), the results must be regarded as highly questionable. A slightly later paper by Alty and Nicoll (9) somewhat improved the experimental arrangements and examined carbon tetrachloride as well as water. The results agreed with the previous investigation for water, but $\sigma$ for carbon tetrachloride was found to be unity. A brief note by Alty (4) in 1932 indicated that a new and considerably improved experimental method would be reported soon.

Papers by Alty (5) and by Alty and McKay (8) describe this later work in detail. The experimental procedure previously used was abandoned, and a rather ingenious idea was introduced to provide a more accurate measure of the surface temperature. This was to allow the evaporating liquid to form drops on a specially prepared capillary tip. The weight of these drops at the time of their fall is a function of the geometry of the tip and the surface tension of the liquid, and this

## 42 THE SIMPLE THEORY OF INTERPHASE MASS TRANSFER

property is presumably a function of the surface temperature. Therefore, if all these functional relationships are known, a unique method is available for determining the surface temperature of a nonisothermal liquid. While certain minor objections to this procedure might be raised, it must be admitted that it was an intriguing idea which avoided the use of any foreign object, such as a thermocouple, at the liquid surface. There may be some question, however, as to the accuracy possible by this method.[4]

Unfortunately, this new method of surface temperature determination introduces a new complication. Vaporization no longer is a steady-state process. The equation which Alty used to interpret these later data was essentially (3.2-4), which he wrote in differential form as

$$d\bar{m} = \sigma \sqrt{\frac{M}{2\pi R T_s}} (P_s^* - P_0) A d\theta , \qquad (3.4\text{-}1)$$

where $d\bar{m}$ is the mass of water evaporated in time $d\theta$ through a surface of area $A$. The other symbols have their usual meaning. Alty integrates (3.4-1) to obtain

$$\bar{m} = \sigma \sqrt{\frac{M}{2\pi R T_s}} (P_s^* - P_0) \int_0^\theta A d\theta , \qquad (3.4\text{-}2)$$

where $\bar{m}$ is the total mass of water evaporated during the time, $\theta$, when the drop was forming. It will be noted that $P_s^*$ and $T_s$ have been considered constants although the liquid which forms the drop is at a higher temperature initially than it is when the drop falls. If an equation similar to (3.4-2) can be used at all, it would seem necessary to use some sort of appropriate average values of $P_s^*$ and $T_s$ which would be intermediate between their initial and final values.

A much more serious objection to Alty's interpretation, however, arises from a question as to the applicability of (3.2-4) to the physical situation existing in his apparatus. If it is assumed that the gas temperature, $T_0$, is equal to the liquid temperature, $T_s$, at the instant the drop falls, then it certainly cannot be equal to it at the time the next drop begins to form. It would be interesting to interpret the data in terms of (3.2-3); but, unfortunately, $T_s$ is known only at the instants before and after the drop falls, and no effort was made to measure $T_0$ at any time. These considerations would seem to indicate that the significance of Alty's values for $\sigma$ is very questionable. Unfortunately, the complications introduced by the unsteady-state nature of these experiments outweigh the appealing feature of Alty's method of surface temperature measurement.

---

[4]See Prüger, reference 71, p. 225, on this point.

## THE SIMPLE THEORY OF INTERPHASE MASS TRANSFER

In the experimental method which was used by Prüger (71), a small quantity of liquid was evaporated from a flat horizontal surface at an approximately constant rate. The experiments were made at atmospheric pressure. Two fine thermocouples (approximately 0.04 mm in diameter) were located above and below the liquid surface, and, as its level dropped because of vaporization, the lower thermocouple (which was stationary with respect to the apparatus) traversed the region in the liquid near the phase interface. The distance between it and the surface was accurately observed by a microscope. This distance, the difference in the temperatures of the two thermocouples, and the rate of evaporation provided Prüger with the information he felt was necessary to determine $\sigma$ in (3.2-4). The temperature difference was plotted as a function of distance from the phase interface and was found to give a curve which was linear except where the thermocouple broke through the surface. Deviations were obviously to be expected here, so Prüger extrapolated the linear portion of the curve to the interface and took this as the interfacial temperature difference.

From this temperature difference he calculated the pressure difference which appears in (3.2-4) by the equation

$$P_s^* - P_0 = \frac{dP^*}{dT}(T_s - T_0), \qquad (3.4\text{-}3)$$

where $dP^*/dT$ is the slope of the vapor pressure-temperature curve. This calculation is acceptable only if $P_0$ and $T_0$ are also on the equilibrium curve. Prüger evidently subscribed to this idea in his original paper, but his later theoretical contribution (72) expresses doubt on this point. The writer can see no justification for this supposition; if it were correct, then the form of (3.2-4) would certainly seem to be in error, since it makes no allowance for interfacial temperature differences.

For water and carbon tetrachloride vaporizing at atmospheric pressure, Prüger reports interfacial temperature differences ranging from several hundredths to a thousandth of a degree centigrade. According to his interpretation of the data, this corresponds to $\sigma = 0.04$ for water and unity for carbon tetrachloride. These figures are based on averages of a considerable number of experiments, and the statistical interpretation of the data is perhaps open to serious question. While no great significance can be given to Prüger's results, it is felt that his method would yield valuable information if the experiments were repeated at lower pressures, where effects of greater magnitude would be expected. (This procedure might require examination of less volatile substances.) It was learned from a former colleague (97) of Dr. Prüger that the latter had been unable to make any further experiments along these lines before his death in the last war.

# CHAPTER IV

# INTERPHASE MASS TRANSFER IN MULTICOMPONENT SYSTEMS

**4.1. The Absolute Rate of Vaporization for a Component of a Multicomponent Liquid or Solid.** The extension of the theory for the absolute rate of vaporization of a pure substance to multicomponent systems involves no new ideas. Again, it is assumed that for systems in equilibrium the gas phase exists as a uniform gas just at the phase interface. From the velocity distribution of each component in this uniform gas, the absolute rate of vaporization is deduced. Thus, for the component $A$ the absolute rate of vaporization is given by $\sigma_A w_{A_{s+}}$, where

$$w_{A_{s+}} = \frac{\tilde{\gamma}_{A_s}}{2\beta_{A_s} \pi^{1/2}} \tag{4.1-1a}$$

$$= \tilde{\gamma}_{A_s} \sqrt{\frac{RT_s}{2\pi M_A}} \tag{4.1-1b}$$

$$= \tilde{P}_{A_s} \sqrt{\frac{M_A}{2\pi RT_s}}. \tag{4.1-1c}$$

The condensation coefficient for component $A$, $\sigma_A$, retains its interpretation as being independent of the vector velocity of incidence or emission of molecules. It is dependent only on the state of the liquid or solid surface and can be expressed as a function of its temperature, pressure, and composition.

The tildes on the symbols for density and pressure in (4.1-1) denote the fact that these represent the properties of a gas which is in equilibrium with the liquid or solid surface. A complete criterion for thermodynamic equilibrium between two multicomponent phases may be expressed as requiring that their temperatures and the fugacities of any component in each of them be the same. For a system of two phases at the same temperature, the fugacity of component $A$ in the condensed phase may be given as a function of its composition and pressure:

$$f_{A_s}' = f_{A_s}'(P_s, x_{A_s}, x_{B_s}, \dots),$$

where the $x$'s are parameters which define the phase's composition. A similar relation will be true for the gas phase:

$$f_{A_s}{}'' = f_{A_s}{}''(P_s{}'', y_{A_s}, y_{B_s}, \ldots).$$

If $f' = f''$ for every component, it follows that $P_s'' = \bar{P}_s$, the pressure of a gas phase in equilibrium with the liquid or solid. If all the properties of the surface are given, it is evident that there will be $n$ independent equations obtained by equating the fugacities of the components with $n - 1$ independent composition parameters and the pressure unknown for the gas phase. It is clear then that there is a gas phase which can be in equilibrium with any specified condition of the liquid or solid surface.

Unfortunately, the form of the above fugacity equations is generally unknown. Experimental vapor-liquid or vapor-solid data are available only for a condensed phase which exists under its own vapor pressure. Therefore, we are forced to make the approximations $\bar{P}_s = P_s^*$ and

$$\bar{y}_{A_s} = y_{A_s}^*.$$

Since the pressure on a liquid or solid surface will ordinarily deviate very little from its vapor pressure, this approximation will be quite satisfactory. Then (4.1-1) becomes

$$w_{A_{s+}} = \frac{y_{A_s}^*}{2\beta_{A_s} \pi^{1/2}} \qquad (4.1\text{-}2a)$$

$$= y_{A_s}^* \sqrt{\frac{RT_s}{2\pi M_A}} \qquad (4.1\text{-}2b)$$

$$= P_{A_s}^* \sqrt{\frac{M_A}{2\pi RT_s}}. \qquad (4.1\text{-}2c)$$

It is convenient in discussing multicomponent systems to speak of the molal rates of flow which will be denoted by $\omega$, the mol fractions $y_A$, $y_B$, etc., and the molal density, $\rho$. Then (4.1-2) may be written

$$\omega_{A_{s+}} = \frac{y_{A_s}^* \rho_s^*}{2\beta_{A_s} \pi^{1/2}} \qquad (4.1\text{-}3a)$$

$$= y_{A_s}{}^*P_s{}^* \sqrt{\frac{RT_s}{2\pi M_A}} \qquad (4.1\text{-}3b)$$

$$= y_{A_s}{}^*P_s{}^* \sqrt{\frac{1}{2\pi M_A RT_s}}. \qquad (4.1\text{-}3c)$$

The absolute rates of vaporization of the components of a multicomponent liquid or solid are expected to be independent of the nature of the gas phase above them for the same reasons which were given for the case of the pure substance (see Section 2.2).

**4.2. The Absolute Rate of Condensation for a Component of a Multicomponent Gas under Nonequilibrium Conditions.** The calculation of $w_{A_{0^-}}$, the mass rate of flow of molecules of $A$ toward the phase interface, involves a complication which did not exist in the case of the pure substance. Except under very special circumstances, the rates of transfer will not be the same for all components in the gas. Therefore, if it is desired to approximate the velocity distribution for each component by (1.2-9a), a question arises as to the mean velocity of the gas. This will be taken as the molal linear velocity of the gas, that is, the vector sum of the molal rates of transfer of each of the components divided by the total molal density. This is consistent with the simple theory of diffusion.[1]

The distribution function for the component $A$ at the phase interface is then

$$f_{A_0} = n_{A_0} \left(\frac{\beta_{A_0}}{\pi^{1/2}}\right)^3 e^{-\beta_{A_0}^2[(U_A - U_{m_0})^2 + V_A^2 + W_A^2]}, \qquad (4.2\text{-}1)$$

where $U_{m_0}$ is the molal linear velocity of the gas and the subscript 0 denotes the properties of the gas phase at the phase interface. The molal rate of flow of $A$ to the interface may then be calculated as

$$\omega_{A_{0^-}} = \int_{-\infty}^{\infty} \int_{-\infty}^{\infty} \int_{-\infty}^{0} \frac{m_A U_A}{M_A} f_{A_0} dU_A dV_A dW_A. \qquad (4.2\text{-}2)$$

If (4.2-2) is compared with (3.1-1), it may immediately be written

$$\omega_{A_{0^-}} = -\left(\frac{\rho_{A_0}}{\rho_{A_s}{}^*}\right) \left(\frac{T_0}{T_s}\right)^{1/2} \Gamma_A \omega_{A_{s+}} \qquad (4.2\text{-}3)$$

---

[1] See reference 39, p. 184 ff.

or

$$\omega_{A_{0-}} = -\left(\frac{y_{A_0}}{y_{A_s}^*}\right)\left(\frac{\rho_0}{\rho_s^*}\right)\left(\frac{T_0}{T_s}\right)^{1/2} \Gamma_A \omega_{A_{s+}}, \quad (4.2\text{-}4)$$

where

$$\Gamma_A = e^{-\beta_{A_0}^2 U_{m_0}^2} - \beta_{A_0} U_{m_0} \pi^{1/2}[1 - \Phi(\beta_{A_0} U_{m_0})]. \quad (4.2\text{-}5)$$

We may now define, and express in terms of the dimensionless ratios appearing in (4.2-4),

$$\phi_{A_0} = \beta_{A_0} U_{m_0}$$

$$= \frac{y_{A_0}}{2\pi^{1/2}}\left(\frac{2\pi^{1/2}\beta_{A_s}}{y_{A_s}^*\rho_s^*}\right)\left(\frac{\beta_{A_0}}{\beta_{A_s}}\right)\left(\frac{y_{A_0}}{y_{A_s}^*}\right)^{-1}\left(\frac{\rho_0}{\rho_s^*}\right)^{-1} U_{m_0}\rho_0, \quad (4.2\text{-}6)$$

and noting (4.1-3a) and (1.2-8), we have

$$\phi_{A_0} = \frac{y_{A_0}}{2\pi^{1/2}}\left(\frac{y_{A_0}}{y_{A_s}^*}\right)^{-1}\left(\frac{\rho_0}{\rho_s^*}\right)^{-1}\left(\frac{T_0}{T_s}\right)^{-1/2} \frac{\omega}{\omega_{A_{s+}}}. \quad (4.2\text{-}7)$$

When departures from equilibrium are small, an approximate equation may be used to calculate $\phi_{A_0}$:

$$\phi_{A_0} = \frac{y_{A_0}}{2\pi^{1/2}} \frac{\omega}{\omega_{A_{s+}}}. \quad (4.2\text{-}7a)$$

The curves in figure 3 or the results tabulated in Appendix B can be used to calculate $\Gamma_A$ once $\phi_{A_0}$ is known.

The absolute (molal) rate of condensation of component $A$ is simply

$$\sigma_A \omega_{A_{0-}},$$

where $\sigma_A$ is the condensation coefficient for component $A$. Entirely analogous equations to all of those developed above are true for the other components of the multicomponent gas.

**4.3. Interphase Mass Transfer in a Binary System.** For simplification, the discussion will now be restricted to a binary system, although all of the equations deduced in the preceding two sections are true for a system of any number of components. The net molal rate of transfer for the component $A$ is given by the sum of its absolute (molal) rates of vaporization and condensation:

$$\omega_A = \sigma_A \omega_{A_{s+}} - \sigma_A \left(\frac{y_{A_0}}{y_{A_s}^*}\right)\left(\frac{\rho_0}{\rho_s^*}\right)\left(\frac{T_0}{T_s}\right)^{1/2} \Gamma_A \omega_{A_{s+}}, \quad (4.3\text{-}1)$$

$$\frac{\omega_A}{\omega_{A_{s+}}} = \sigma_A \left[1 - \left(\frac{y_{A_0}}{y_{A_s}^*}\right)\left(\frac{\rho_0}{\rho_s^*}\right)\left(\frac{T_0}{T_s}\right)^{1/2} \Gamma_A\right]. \quad (4.3\text{-}2a)$$

Similarly, for the component $B$ (noting that $y_B = 1 - y_A$) we have

$$\frac{\omega_B}{\omega_{B_{s+}}} = \sigma_B \left[1 - \left(\frac{1-y_{A_0}}{1-y_{A_s}^*}\right)\left(\frac{\rho_0}{\rho_s^*}\right)\left(\frac{T_0}{T_s}\right)^{1/2} \Gamma_B\right]. \quad (4.3\text{-}2b)$$

Solving (4.3-2) for the density and temperature ratios gives

$$-\left(\frac{\rho_0}{\rho_s^*}\right)\left(\frac{T_0}{T_s}\right)^{1/2} = \left[\frac{\omega_A}{\sigma_A \omega_{A_{s+}}} - 1\right]\left[\left(\frac{y_{A_0}}{y_{A_s}^*}\right)\Gamma_A\right]^{-1} \quad (4.3\text{-}3a)$$

$$= \left[\frac{\omega_B}{\sigma_B \omega_{B_{s+}}} - 1\right]\left[\left(\frac{1-y_{A_0}}{1-y_{A_s}^*}\right)\Gamma_B\right]^{-1}. \quad (4.3\text{-}3b)$$

Equating (4.3-3a) and (4.3-3b) and solving for $y_{A_0}/y_{A_s}^*$, we have

$$\frac{y_{A_0}}{y_{A_s}^*} = \left\{y_{A_s}^* + \frac{[(\omega_B/\sigma_B \omega_{B_{s+}})-1]\Gamma_A}{[(\omega_A/\sigma_A \omega_{A_{s+}})-1]\Gamma_B}(1-y_{A_s}^*)\right\}^{-1}. \quad (4.3\text{-}4)$$

There are two special cases of relative flow of components $A$ and $B$ which are of particular interest. For equimolal countertransfer of $A$ and $B$, the molal linear velocity of the gas is zero and the factors $\Gamma_A = \Gamma_B = 1$. Then (4.3-4) becomes

$$\frac{y_{A_0}}{y_{A_s}^*} = \left\{ y_{A_s}^* + \frac{[(\omega_B/\sigma_B \omega_{B_{s+}}) - 1]}{[(\omega_A/\sigma_A \omega_{A_{s+}}) - 1]} (1 - y_{A_s}^*) \right\}^{-1} . \quad (4.3\text{-}4a)$$

For the case of mass transfer only of component $A$, we have $\omega_B = 0$ and (4.3-4) becomes

$$\frac{y_{A_0}}{y_{A_s}^*} = \left\{ y_{A_s}^* + \frac{(1 - y_{A_s}^*)\Gamma_A}{(1 - \omega_A/\sigma_A \omega_{A_{s+}})\Gamma_B} \right\}^{-1} . \quad (4.3\text{-}4b)$$

It is interesting to observe that (4.3-4) permits calculation of

$$y_{A_0}/y_{A_s}^*$$

without any necessity that the temperature or density ratios which were eliminated from (4.3-3) be known. Since these eliminated terms are not of particular importance in intraphase mass transfer calculations, this is most convenient.[2]

There are a few references in the literature which deal with experimental investigation of interfacial resistances to mass transfer in systems of more than one component. Papers by Miyamoto (64, 65) and Uhara (92) deal most directly with the problem, but are unfortunately complicated by chemical reactions and diffusional effects, so that no very useful information can be gained from them.[3] A method used by Brookfield *et al.* (18) for determining gas-phase diffusivities was suggested as a possible means of also determining the condensation coefficient for a pure liquid evaporating into an inert gas. (The theoretical analysis associated with this method was based on the supposition that (3.2-4) applied to the evaporating molecules.) The particular data obtained, however, were not suited to the calculation.

[2] If $(\rho_0/\rho_{s}^* )(T_0/T_s)^{1/2}$ differs considerably from unity, it will be necessary to calculate it from (4.3-3a) or (4.3-3b), because it appears in $\phi_{A_0}$ and $\phi_{B_0}$, which are needed to calculate $\Gamma_A$ and $\Gamma_B$. Separation of the values of the density and temperature ratios is not necessary, however; it would require considerations similar to those in Section 3.3 for the pure substance.

[3] Parts of Miyamoto's theoretical discussion (for example, the concept that a gas molecule requires a minimum velocity to condense) are in contradiction to views expressed earlier in this book.

# CHAPTER V

# ANOTHER THEORY OF INTERPHASE MASS TRANSFER OF A PURE SUBSTANCE

**5.1. The Exact Description of the Gas at the Phase Interface.** When a pure substance evaporates or condenses because of energy supplied or removed through the condensed phase, it is apparent that mass can be transferred across the phase interface without any intraphase process occurring in the gas. It might at first be thought, therefore, that the gas simply drifts to or from the phase interface as a uniform gas. This assumption, however, would require an unreasonable law for the velocity distribution of molecules emitted from the liquid or solid surface. The velocity distribution of these molecules would have to adapt itself to the rate and direction of any interphase mass transfer process which might take place.

In Chapter II the characteristics of molecules emitted from the surface of a condensed phase were formulated as a part of the theory of the absolute rate of vaporization. It is important to recognize at this point that there is no part of that discussion which does not follow mathematically from the set of mutually consistent assumptions which were stated there. It was also noted in Chapter II that the available reliable experimental evidence suggests that the condensation coefficient, $\sigma$, is either unity or close to it. In this chapter it will be assumed that $\sigma$ is unity and that the velocity distribution of molecules emitted from a liquid or solid surface is given by

$$\tilde{f}_s = \tilde{n}_s \left(\frac{\beta_s^3}{\pi^{3/2}}\right) e^{-\beta_s^2 [U^2 + V^2 + W^2]} \tag{5.1-1}$$

This assumption can be true only for an infinitesimally small distance from the phase interface because, as these molecules move out into the gas, they encounter others with a different velocity distribution.

At some distance of several mean free paths from the phase interface, the gas must be moving as a uniform gas, provided no intraphase transfer process is occurring. As was noted in Section 1.2, a nonuniform gas rapidly assumes a normal velocity distribution when disturbing influences are removed. This distribution will be denoted by

$$f_1 = n_1 \left(\frac{\beta_1^3}{\pi^{3/2}}\right) e^{-\beta_1^2[(U-U_1)^2 + V^2 + W^2]} \qquad (5.1\text{-}2)$$

In Chapters III and IV, the rate at which mass flows from the liquid or solid surface was correctly computed from (5.1-1) (although allowance was made for $\sigma \neq 1$). The absolute rate of condensation was, however, based on (5.1-2), and this cannot possibly be a rigorous description of those molecules with $U < 0$ just at the phase interface. Therefore, the results of Chapters III and IV are approximate. This chapter attempts to develop an interphase mass transfer theory in which the description of molecular behavior is mathematically consistent with the theory of the absolute rate of vaporization developed in Chapter II. The results of this theory will then be compared with those of Chapter III.

The nucleus of the present treatment is the adoption of a modified distribution function for the gas at the phase interface. The theory is first developed for monatomic molecules and is then extended to polyatomic molecules in a more approximate manner. Only one other investigator (as far as is known) has discussed the problem of interphase mass transfer of a pure substance from the viewpoint of a modified distribution function, and, before we proceed with the present analysis, his contribution will be reviewed.

**5.2. The Crout Theory of the Evaporation of a Pure Substance.** Two papers were published in 1936 by Crout (23) and by Giddings and Crout (29) on the problem of evaporation of monatomic and polyatomic gases from liquids or solids. These papers do not appear to have been discussed in the literature since their publication, and the authors have not compared their theory with earlier references on the subject.

It is assumed that a Maxwell velocity distribution exists in the gas near the phase interface when a two-phase system is in equilibrium. Crout then demonstrates that it is impossible to explain exactly the behavior of an evaporating gas in terms of this or any other single Maxwell distribution function. It is then assumed that during any evaporation process the velocity distribution of all the gas molecules just at the phase interface has the form

$$f_0 = A e^{-m[h_L(U-U_0)^2 + h_T V^2 + h_T W^2]}, \qquad (5.2\text{-}1)$$

where $A$, $h_L$, and $h_T$ are constants fixed by the conditions of the problem. The constant $A$ is determined from

$$n_0 = \int_{-\infty}^{\infty} \int_{-\infty}^{\infty} \int_{-\infty}^{\infty} f_0 \, dU \, dV \, dW, \qquad (5.2\text{-}2)$$

so that (5.2-1) may be written

$$f_0 = n_0 \left( \frac{h_L h_T^2 m}{\pi^3} \right)^{1/2} e^{-m[h_L(U-U_0)^2 + h_T V^2 + h_T W^2]}. \qquad (5.2\text{-}3)$$

The four constants in the above equation ($n_0$, $h_L$, $h_T$, and $U_0$) are evaluated by calculating the rates of emission of mass, momentum, and energy from the liquid or solid surface according to the equilibrium Maxwell velocity distribution function and then equating these to the same rates calculated for those molecules with $U > 0$ from the distribution function (5.2-3). These three equations, plus the fact that the (net) rate of evaporation must take a specified value, determine the four unknowns in (5.2-3) and completely describe the gas at the interface.

For the same reasons discussed in Section 5.1, Crout considers that eventually the evaporating gas becomes uniform and has the distribution function (5.1-2). The three conditions which the gas at the interface imposes on this uniform gas [1] are then used to calculate its properties.

The procedure which has been used to calculate the constants in (5.2-3) is rather peculiar. The magnitude of mass, momentum, and energy flow from the liquid or solid surface is calculated on the basis of a Maxwell velocity distribution, but those molecules with $U > 0$ do not have such a distribution, according to (5.2-3). No reason is given for the choice of (5.2-1) as an appropriate form of distribution function, other than the implication that it resembles the Maxwell velocity distribution and contains "a sufficient number of constants, or parameters, to render possible a good approximation to actual conditions." It would appear, however, that the form of distribution function assumed in (5.2-1) is physically inconsistent with the velocity distribution which has been supposed for the molecules emitted from the surface of the condensed phase.

### 5.3. The Assumed Distribution Function for the Gas at the Phase Interface.

Consider 0 and $s$ to be planes taken parallel to the phase interface and just in the gas and the condensed phase, respectively. Let another plane 1 be parallel to the phase interface and located in the gas sufficiently far away that the gas there may be considered uniform. Also, let w and E be vectors representing the flow of mass and

---

[1] These will be discussed in Section 5.3.

energy in the gas, and let $P_s$ be a tensor representing the flow of momentum relative to the phase interface. This tensor is denoted by $P_s$ because it is identical with the surface pressure tensor defined in Section 1.2d, provided the mean velocity of molecules at $s$ is neglected.

The application of classical mechanics to the region between 0 and 1 requires that the divergences of w, E, and $P_s$ be zero; that is,

$$\nabla \cdot \mathbf{w} = \nabla \cdot \mathbf{E} = \nabla \cdot P_s = 0. \qquad (5.3\text{-}1)$$

The only processes which occur in this region are collisions between the molecules which constitute the gas, and neither mass, momentum, nor energy can be created or destroyed as a result of such collisions. The quantities w, E, and $P_s$ are therefore called summational invariants of the gas, and for a system of monatomic molecules without detailed structure these are the only possible independent summational invariants.[2] It is apparent that (5.3-1) requires the values of w, E, and $P_s$ to be the same at planes 0 and 1, so that no subscripts are necessary on these quantities.

The distribution function for those molecules at plane 0 with $U > 0$ is known from (5.1-1). It is necessary to know the distribution function only for those molecules with $U < 0$ in order to specify fully the behavior of the gas at the phase interface. Ideally, this should be done by an analysis of the collision processes between the molecules in the region between 0 and 1, but this would be a problem beyond the scope of the present study.

On the other hand, a reasonable form of distribution function may be assumed for these molecules. Particularly if deviations from equilibrium conditions are not too great, it is not difficult to make a reasonable choice. Furry (28) has shown that this could be done for the case of diffusion in binary systems, as mentioned in Section 1.2c. It will be assumed here that a similar correction factor may be applied to the distribution function for the gas which would be in equilibrium with the liquid or solid surface. Then the distribution function for those molecules at the interface with $U < 0$ is simply

$$f_{0-} = (1 + BU)\bar{f}_s, \qquad (5.3\text{-}2)$$

where $B$ is a constant to be determined. If $T_s$ and the rate of interphase mass transfer are specified, then the distribution function (5.3-2)—and hence the state of the gas—is fully determined, both at planes 0 and 1.

If it had been decided to use a correction factor which had two independent constants, then specifying $T_s$ and $w$ would no longer be

---

[2] See reference 21, pp. 50-51.

54  ANOTHER THEORY OF INTERPHASE MASS TRANSFER

sufficient fully to determine the state of the gas. There would seem to be only two alternative possibilities for a solution in this case. Either another variable, such as the temperature or the pressure of the gas at 1, must be fixed, or another relation must exist between the variables involved in the problem. The first of these possibilities does not seem acceptable, because, once the rate of mass transfer and the temperature of the liquid or solid surface have been specified, it does not appear to be possible to vary independently the condition of the gas phase. The possibility that another relationship exists between the variables in the problem is also doubtful. The three summational invariants used in (5.3-1) are the only ones (for a monatomic gas) which can relate its states at two different points.

**5.4. Transport of Mass, Momentum, and Energy in the Gas at the Phase Interface.** The rate at which any property, $\psi$, of a particular molecule is transported in the direction of the phase interface is $\mathbf{l} \cdot \mathbf{C}\psi$ or $U\psi$. The total rate of transport of the property $\psi$ can easily be obtained from the distribution function for the gas at the phase interface, as given by (5.1-1) and (5.3-2):

$$\sum (U\psi) = \int_{-\infty}^{\infty} \int_{-\infty}^{\infty} \int_{0}^{\infty} (U\psi) \bar{f}_s \, dU \, dV \, dW$$

$$+ \int_{-\infty}^{\infty} \int_{-\infty}^{\infty} \int_{-\infty}^{0} (U\psi)(1 + BU) \bar{f}_s \, dU \, dV \, dW . \qquad (5.4\text{-}1)$$

This may be written in the alternative forms,

$$\sum (U\psi) = \int_{-\infty}^{\infty} \int_{-\infty}^{\infty} \int_{-\infty}^{\infty} (U\psi) \bar{f}_s \, dU \, dV \, dW$$

$$+ B \int_{-\infty}^{\infty} \int_{-\infty}^{\infty} \int_{0}^{\infty} (U^2 \psi) \bar{f}_s \, dU \, dV \, dW \qquad (5.4\text{-}2a)$$

$$= \int_{-\infty}^{\infty} \int_{-\infty}^{\infty} \int_{-\infty}^{\infty} (U\psi) \bar{f}_s \, dU \, dV \, dW$$

$$- B \int_{-\infty}^{\infty} \int_{-\infty}^{\infty} \int_{0}^{\infty} (U^2 \psi) \bar{f}_s \, dU \, dV \, dW . \qquad (5.4\text{-}2b)$$

If $U^2\psi$ is an even power of $U$, then (5.4-2a) should be used; if it is an odd power of $U$, then (5.4-2b) should be used.

The constant $B$ is easily expressed in terms of the rate of interphase mass transfer. In this case, the property, $\psi$, under consideration is simply the mass of the molecule $m$, and thus, from (5.4-2a), we have

$$w = \int_{-\infty}^{\infty} \int_{-\infty}^{\infty} \int_{-\infty}^{\infty} Um\bar{f}_s dUdVdW$$

$$+ B \int_{-\infty}^{\infty} \int_{-\infty}^{\infty} \int_{0}^{\infty} U^2 m\bar{f}_s dUdVdW. \quad (5.4\text{-}3)$$

The first integral of (5.4-3) will be recognized as the rate of mass transport in the stationary uniform gas which would be in equilibrium with the liquid or solid surface. This is zero, so (5.4-3) becomes simply

$$w = B \int_{-\infty}^{\infty} \int_{-\infty}^{\infty} \int_{0}^{\infty} U^2 m \tilde{n}_s \left(\frac{\beta_s^3}{\pi^{3/2}}\right) e^{-\beta_s^2 [U^2 + V^2 + W^2]} dUdVdW \quad (5.4\text{-}4)$$

$$= B\tilde{\gamma}_s \left(\frac{\beta_s^3}{\pi^{3/2}}\right) \int_{-\infty}^{\infty} \int_{-\infty}^{\infty} \int_{0}^{\infty} U^2 e^{-\beta_s^2 [U^2 + V^2 + W^2]} dUdVdW \quad (5.4\text{-}5)$$

$$= B\tilde{\gamma}_s \left(\frac{\beta_s^3}{\pi^{3/2}}\right) \left(\frac{\pi^{1/2}}{4\beta_s^3}\right) \left(\frac{\pi^{1/2}}{\beta_s}\right) \left(\frac{\pi^{1/2}}{\beta_s}\right) \quad (5.4\text{-}6)$$

$$= \frac{B\tilde{\gamma}_s}{4\beta_s^2}. \quad (5.4\text{-}7)$$

Noting the definition of $w_{s+}$ given by (2.1-6a), we obtain the constant $B$ from (5.4-7) as

$$B = \frac{2\beta_s}{\pi^{1/2}} \frac{w}{w_{s+}} \quad (5.4\text{-}8)$$

The density of the gas just at the phase interface may now easily be computed from

$$n_0 = \int_{-\infty}^{\infty} \int_{-\infty}^{\infty} \int_{-\infty}^{\infty} \tilde{f}_s \, dU dV dW - B \int_{-\infty}^{\infty} \int_{-\infty}^{\infty} \int_{0}^{\infty} U \tilde{f}_s \, dU dV dW \quad (5.4\text{-}9)$$

$$= \tilde{n}_s - B \tilde{n}_s \left( \frac{\beta_s^3}{\pi^{3/2}} \right) \int_{-\infty}^{\infty} \int_{-\infty}^{\infty} \int_{0}^{\infty} U e^{-\beta_s^2 [U^2 + V^2 + W^2]} dU dV dW \quad (5.4\text{-}10)$$

$$= \tilde{n}_s - B \tilde{n}_s \left( \frac{\beta_s^3}{\pi^{3/2}} \right) \left( \frac{1}{2\beta_s^2} \right) \left( \frac{\pi^{1/2}}{\beta_s} \right) \left( \frac{\pi^{1/2}}{\beta_s} \right). \quad (5.4\text{-}11)$$

Substituting for $B$ its value from (5.4-8) and simplifying, we obtain

$$n_0 = \tilde{n}_s \left[ 1 - \frac{1}{\pi} \frac{w}{w_{s+}} \right] \quad (5.4\text{-}12)$$

or, in dimensionless form

$$\frac{n_0}{\tilde{n}_s} = \frac{\gamma_0}{\tilde{\gamma}_s} = 1 - \frac{1}{\pi} \frac{w}{w_{s+}}. \quad (5.4\text{-}13)$$

If $\psi$ is the $x$ component of the momentum of a molecule, $\mathbf{i} \cdot m\mathbf{C} = mU$, then the total rate of transport of this component in the direction normal to the phase interface is, by (5.4-2b),

$$\mathbf{i} \cdot \mathsf{P}_s \cdot \mathbf{i} = \int_{-\infty}^{\infty} \int_{-\infty}^{\infty} \int_{-\infty}^{\infty} U^2 m \tilde{f}_s \, dU dV dW$$

$$- B \int_{-\infty}^{\infty} \int_{-\infty}^{\infty} \int_{0}^{\infty} U^3 m \tilde{f}_s \, dU dV dW. \quad (5.4\text{-}14)$$

The first integral of (5.4-14) will be recognized from (1.2-3) as the magnitude of the ii component of the pressure tensor of the uniform gas which would be in equilibrium with the liquid or solid surface (see (1.2-16) and (1.2-20)), and therefore (5.4-14) becomes

$$\mathbf{i} \cdot \mathsf{P}_s \cdot \mathbf{i} = \frac{\tilde{\gamma}_s}{2\beta_s^2} - B \int_{-\infty}^{\infty} \int_{-\infty}^{\infty} \int_{0}^{\infty} U^3 m \tilde{n}_s \left( \frac{\beta_s^3}{\pi^{3/2}} \right) e^{-\beta_s^2 [U^2 + V^2 + W^2]}$$

$$\times dU dV dW \quad (5.4\text{-}15)$$

$$= \frac{\tilde{\gamma}_s}{2\beta_s{}^2} - B\tilde{\gamma}_s \left(\frac{\beta_s{}^3}{\pi^{3/2}}\right) \int_{-\infty}^{\infty} \int_{-\infty}^{\infty} \int_0^{\infty} U^3 e^{-\beta_s{}^2[U^2+V^2+W^2]} dUdVdW \quad (5.4\text{-}16)$$

$$= \frac{\tilde{\gamma}_s}{2\beta_s{}^2} - B\tilde{\gamma}_s \left(\frac{\beta_s{}^3}{\pi^{3/2}}\right)\left(\frac{1}{2\beta_s{}^4}\right)\left(\frac{\pi^{1/2}}{\beta_s}\right)\left(\frac{\pi^{1/2}}{\beta_s}\right). \quad (5.4\text{-}17)$$

Substituting $B$ from (5.4-8), we obtain

$$\mathbf{i} \cdot \mathsf{P}_s \cdot \mathbf{i} = \frac{\tilde{\gamma}_s}{2\beta_s{}^2} - \frac{\tilde{\gamma}_s}{\pi\beta_s{}^2}\frac{w}{w_{s+}} \quad (5.4\text{-}18)$$

$$= \frac{\tilde{\gamma}_s}{2\beta_s{}^2}\left[1 - \frac{2}{\pi}\frac{w}{w_{s+}}\right]. \quad (5.4\text{-}19)$$

Since $\mathbf{i} \cdot \mathsf{P}_s \cdot \mathbf{i}$ is the total rate of transport of the $x$ component of momentum in the $x$ direction, it is also equal to the magnitude of the ii component of the surface pressure tensor defined in (1.2-18) (neglecting the mean velocity of the liquid or solid molecules). Therefore, it also gives the magnitude of the normal component of the surface pressure, $P_{si}$. This may be expressed in terms of the equilibrium vapor pressure, $P_s^*$, by noting (1.2-20):

$$P_{si} = P_s^* \left[1 - \frac{2}{\pi}\frac{w}{w_{s+}}\right]. \quad (5.4\text{-}20)$$

Equation (5.4-20) permits the use of (2.1-1) or (2.1-2) to determine $\tilde{P}_s$ or $\tilde{\gamma}_s$ as distinguished from $P_s^*$ or $\gamma_s^*$. Most practical instances will not warrant this refinement, however.

Since the energy of monatomic molecules is simply their energy of translation, $\tfrac{1}{2}mC^2$, the rate of energy transport in the gas at the phase interface is, from (5.4-2a),

$$E = \int_{-\infty}^{\infty} \int_{-\infty}^{\infty} \int_{-\infty}^{\infty} U(\tfrac{1}{2}mC^2)\tilde{f}_s dUdVdW$$

$$+ B \int_{-\infty}^{\infty} \int_{-\infty}^{\infty} \int_0^{\infty} U^2(\tfrac{1}{2}mC^2)\tilde{f}_s dUdVdW. \quad (5.4\text{-}21)$$

## 58 ANOTHER THEORY OF INTERPHASE MASS TRANSFER

The first integral of (5.4-21) is zero, because there is no energy transport in a stationary uniform gas; therefore, (5.4-21) becomes

$$E = B \int_{-\infty}^{\infty} \int_{-\infty}^{\infty} \int_{0}^{\infty} U^2(U^2 + V^2 + W^2)(\tfrac{1}{2}m)\bar{n}_s \\ \times \left(\frac{\beta_s^3}{\pi^{3/2}}\right) e^{-\beta_s^2[U^2+V^2+W^2]} dU dV dW \tag{5.4-22}$$

$$= \frac{B\tilde{\gamma}_s}{2}\left(\frac{\beta_s^3}{\pi^{3/2}}\right) \int_{-\infty}^{\infty} \int_{-\infty}^{\infty} \int_{0}^{\infty} U^2(U^2 + V^2 + W^2) \\ \times e^{-\beta_s^2[U^2+V^2+W^2]} dU dV dW . \tag{5.4-23}$$

The integral in (5.4-23) is conveniently divided into three parts:

$$\int_{-\infty}^{\infty} \int_{-\infty}^{\infty} \int_{0}^{\infty} U^4 e^{-\beta_s^2[U^2+V^2+W^2]} dU dV dW = \left(\frac{3\pi^{1/2}}{8\beta_s^5}\right)\left(\frac{\pi^{1/2}}{\beta_s}\right)\left(\frac{\pi^{1/2}}{\beta_s}\right), \tag{5.4-24a}$$

$$\int_{-\infty}^{\infty} \int_{-\infty}^{\infty} \int_{0}^{\infty} U^2 V^2 e^{-\beta_s^2[U^2+V^2+W^2]} dU dV dW = \left(\frac{\pi^{1/2}}{4\beta_s^3}\right)\left(\frac{2\pi^{1/2}}{4\beta_s^3}\right)\left(\frac{\pi^{1/2}}{\beta_s}\right), \tag{5.4-24b}$$

$$\int_{-\infty}^{\infty} \int_{-\infty}^{\infty} \int_{0}^{\infty} U^2 W^2 e^{-\beta_s^2[U^2+V^2+W^2]} dU dV dW = \left(\frac{\pi^{1/2}}{4\beta_s^3}\right)\left(\frac{\pi^{1/2}}{\beta_s}\right)\left(\frac{2\pi^{1/2}}{4\beta_s^3}\right). \tag{5.4-24c}$$

Adding the three parts of (5.4-24) and substituting in (5.4-23) gives

$$E = \frac{B\tilde{\gamma}_s}{2}\left(\frac{\beta_s^3}{\pi^{1/2}}\right)\left[\frac{3\pi^{1/2}}{8\beta_s^7} + \frac{2\pi^{1/2}}{8\beta_s^7}\right], \tag{5.4-25}$$

and from (5.4-8), we have

$$E = \frac{5\tilde{\gamma}_s}{8\beta_s{}^3\pi^{1/2}} \frac{w}{w_{s+}} \, . \tag{5.4-26}$$

**5.5. Transport of Mass, Momentum, and Energy in the Uniform Gas.** The distribution function for the uniform gas several mean free paths from the phase interface is given by (5.1-2). The rate of mass transfer in this gas is simply

$$w = \gamma_1 U_1 \, . \tag{5.5-1}$$

The rate at which the $x$ component of momentum is transported in the direction normal to the phase interface is

$$\mathbf{i} \cdot \mathbf{P}_s \cdot \mathbf{i} = \int_{-\infty}^{\infty} \int_{-\infty}^{\infty} \int_{-\infty}^{\infty} mU^2 f_1 dU dV dW \, , \tag{5.5-2}$$

which may also be written

$$\mathbf{i} \cdot \mathbf{P}_s \cdot \mathbf{i} = \int_{-\infty}^{\infty} \int_{-\infty}^{\infty} \int_{-\infty}^{\infty} m(U - U_1)^2 f_1 dU dV dW$$

$$+ \int_{-\infty}^{\infty} \int_{-\infty}^{\infty} \int_{-\infty}^{\infty} 2mUU_1 f_1 dU dV dW$$

$$- \int_{-\infty}^{\infty} \int_{-\infty}^{\infty} \int_{-\infty}^{\infty} mU_1{}^2 f_1 dU dV dW \, . \tag{5.5-3}$$

The first of the integrals in (5.5-3) is the magnitude of the $\mathbf{i}\mathbf{i}$ component of the pressure tensor for the uniform gas at 1. Its value is given by (1.2-20) as

$$\int_{-\infty}^{\infty} \int_{-\infty}^{\infty} \int_{-\infty}^{\infty} m(U - U_1)^2 f_1 dU dV dW = \frac{\gamma_1}{2\beta_1{}^2} \, . \tag{5.5-4}$$

The second and third integrals of (5.5-3) can be expressed in terms of the mean velocity of the gas at 1:

$$\int_{-\infty}^{\infty} \int_{-\infty}^{\infty} \int_{-\infty}^{\infty} 2mU_1 U f_1 dU dV dW = 2\gamma_1 U_1{}^2 \, , \tag{5.5-5}$$

60  ANOTHER THEORY OF INTERPHASE MASS TRANSFER

$$\int_{-\infty}^{\infty}\int_{-\infty}^{\infty}\int_{-\infty}^{\infty} mU_1{}^2 f_1 dU dV dW = \gamma_1 U_1{}^2 . \qquad (5.5\text{-}6)$$

Substituting (5.5-4), (5.5-5), and (5.5-6) in (5.5-3), we obtain

$$\mathbf{i} \cdot \mathbf{P}_s \cdot \mathbf{i} = \frac{\gamma_1}{2\beta_1{}^2}\left[1 + 2\beta_1{}^2 U_1{}^2\right] . \qquad (5.5\text{-}7)$$

The rate of energy transport at 1 is calculated from

$$E = \int_{-\infty}^{\infty}\int_{-\infty}^{\infty}\int_{-\infty}^{\infty} U(\tfrac{1}{2}mC^2) f_1 dU dV dW . \qquad (5.5\text{-}8)$$

The integration of (5.5-8) is straightforward, although the algebra is somewhat involved. It is given in Appendix D, and the result is

$$E = \frac{\gamma_1 \beta_1 U_1}{2\beta_1{}^3}\left[\frac{5}{2} + \beta_1{}^2 U_1{}^2\right] . \qquad (5.5\text{-}9)$$

**5.6. Condition of the Uniform Gas at Plane 1 as a Function of the Rate of Interphase Mass Transfer.** In Section 5.3 it was explained that the rates of mass, momentum, and energy transport must be the same at 1 and at the phase interface. Therefore, equating the momentum equations (5.4-19) and (5.5-7), we have

$$\frac{\gamma_1}{2\beta_1{}^2}\left[1 + 2\beta_1{}^2 U_1{}^2\right] = \frac{\tilde{\gamma}_s}{2\beta_s{}^2}\left[1 - \frac{2}{\pi}\frac{w}{w_{s+}}\right] . \qquad (5.6\text{-}1)$$

As will be recalled from the simple theory of interphase mass transfer (Chapter III), the group $\beta_1 U_1$ may be defined as

$$\phi_1 = \beta_1 U_1 = \frac{1}{2\pi^{1/2}}\left(\frac{\gamma_1}{\gamma_s}\right)^{-1}\left(\frac{T_1}{T_s}\right)^{-1/2}\frac{w}{w_{s+}} . \qquad (5.6\text{-}2)$$

Since $\beta$ is inversely proportional to the square root of the absolute temperature (1.2-8), equation (5.6-1) can be rearranged in the dimensionless form

## ANOTHER THEORY OF INTERPHASE MASS TRANSFER

$$\left(\frac{\gamma_1}{\tilde{\gamma}_s}\right)\left(\frac{T_1}{T_s}\right) = \frac{1 - (2/\pi)w/w_{s+}}{1 + 2\phi_1^2}. \tag{5.6-3}$$

From the energy equations (5.4-26) and (5.5-9), we have

$$\frac{5}{8}\frac{\tilde{\gamma}_s}{\beta_s^3 \pi^{1/2}}\frac{w}{w_{s+}} = \frac{\gamma_1 \beta_1 U_1}{2\beta_1^3}\left[\frac{5}{2} + \beta_1^2 U_1^2\right] \tag{5.6-4}$$

or

$$\frac{5}{4\pi^{1/2}}\left(\frac{\gamma_1}{\tilde{\gamma}_s}\right)^{-1}\left(\frac{T_1}{T_s}\right)^{-3/2}\frac{w}{w_{s+}} = \beta_1 U_1\left[\frac{5}{2} + \beta_1^2 U_1^2\right]. \tag{5.6-5}$$

Replacing $\beta_1 U_1$ with (5.6-2) and denoting $\beta_1^2 U_1^2$ by $\phi_1^2$, we obtain

$$\frac{5}{4\pi^{1/2}}\left(\frac{\gamma_1}{\tilde{\gamma}_s}\right)^{-1}\left(\frac{T_1}{T_s}\right)^{-3/2}\frac{w}{w_{s+}} = \frac{1}{2\pi^{1/2}}\frac{w}{w_{s+}}\left(\frac{T_1}{T_s}\right)^{-1/2}\left(\frac{\gamma_1}{\tilde{\gamma}_s}\right)^{-1}\left[\frac{5}{2} + \phi_1^2\right], \tag{5.6-6}$$

$$\frac{5}{2}\left(\frac{T_1}{T_s}\right)^{-1} = \frac{5}{2} + \phi_1^2, \tag{5.6-7}$$

$$\frac{T_1}{T_s} = \left[1 + \frac{2}{5}\phi_1^2\right]^{-1}. \tag{5.6-8}$$

Substituting (5.6-8) in (5.6-3) gives

$$\frac{\gamma_1}{\tilde{\gamma}_s} = \left[1 - \frac{2}{\pi}\frac{w}{w_{s+}}\right]\frac{1 + 2\phi_1^2/5}{1 + 2\phi_1^2}. \tag{5.6-9}$$

From (5.6-2) and (5.6-8) $\phi_1$ may be written

$$\phi_1 = \frac{1}{2\pi^{1/2}}\left(\frac{\gamma_1}{\tilde{\gamma}_s}\right)^{-1}\left[1 + \frac{2}{5}\phi_1^2\right]^{1/2}\frac{w}{w_{s+}}. \tag{5.6-10}$$

This may be solved for the density ratio to give

$$\frac{\gamma_1}{\bar{\gamma}_s} = \frac{(1 + 2\phi_1^2/5)^{1/2}}{2\pi^{1/2}\phi_1} \frac{w}{w_{s+}}. \qquad (5.6\text{-}11)$$

Eliminating $\gamma_1/\bar{\gamma}_s$ between (5.6-9) and (5.6-11), we obtain $w/w_{s+}$ as an explicit function of $\phi_1$:

$$\frac{w}{w_{s+}} = \left[ \frac{2}{\pi} + \frac{1 + 2\phi_1^2}{(1 + 2\phi_1^2/5)^{1/2}(2\phi_1\pi^{1/2})} \right]^{-1} \qquad (5.6\text{-}12)$$

Therefore, for a given value of the parameter $\phi_1$, the value of $w/w_{s+}$ may be immediately calculated from (5.6-12), and the corresponding values of $T_1/T_s$ and $\gamma_1/\bar{\gamma}_s$ can then be obtained from (5.6-8) and (5.6-9) or (5.6-11), respectively.

**5.7. The Maximum Rate of Interphase Mass Transfer.** Crout (23) has made an interesting observation regarding the maximum rate of evaporation. If the rates of mass, momentum, and energy transport in a uniform gas are specified, then it can be shown that the following equation must be satisfied by $\phi_1$:

$$\frac{\phi_1^2(\phi_1^2 + 5/2)}{(\phi_1^2 + 1/2)^2} = \eta, \qquad (5.7\text{-}1)$$

where $\eta$ is a function of the three transport rates. For an interphase mass transfer process, these three rates have been developed as a function of the rate of interphase mass transfer and the absolute temperature of the liquid or solid surface.

Equation (5.7-1) may be solved for $\phi_1^2$, giving

$$\phi_1^2 = \frac{\eta - (5/2) \pm \sqrt{(25/4) - 4\eta}}{2(1 - \eta)}. \qquad (5.7\text{-}2)$$

It is clear that $\eta$ must be less than or equal to 25/16 for (5.7-2) to have a real solution. Furthermore, for every value of $\eta < 25/16$ there are two corresponding values for $\phi_1^2$. A curve showing $|\phi_1|$ as a function of $\eta$ is reproduced in figure 7 from Crout's calculations.[8] If the state of the evaporating or condensing gas is considered to be a single-valued function of the rate of mass transfer, then only one part of this

---

[3] Crout discusses only the problem of evaporation ($\phi_1 > 0$), but the argument is equally applicable to condensation ($\phi_1 < 0$).

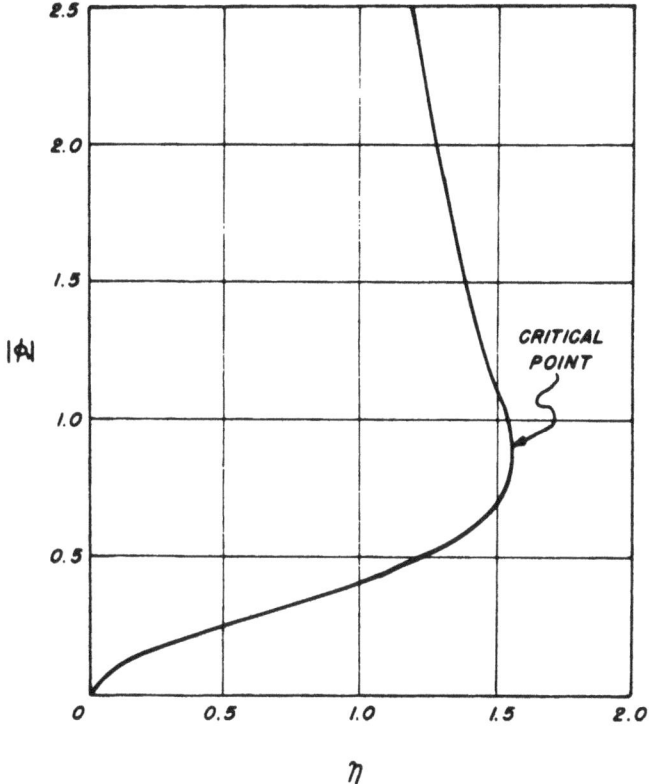

FIGURE 7.

$\phi_1$, as a function of $\eta$ (from Crout, reference 23)

curve can be real. This must be the lower part $(0 \leq |\phi_1| \leq 0.91287...)$, because $\phi_1 = 0$ is certainly a real value.

The value of $w/w_{s+}$ corresponding to the critical value of $\phi_1$ is the maximum rate of interphase mass transfer. As will be apparent in the results given in the next section, it is impossible to exceed this rate in a true interphase process. Although the specific value of $(w/w_{s+})_{max}$ does depend upon the distribution function assumed at the phase interfact, the conclusion that a $(w/w_{s+})_{max}$ does exist is independent of the behavior of the gas at the interface. At the critical point, as Crout has shown, the velocity of the uniform gas is equal to the velocity of sound in the same (stationary) gas. This situation is analogous to the escape of molecules through an orifice from a region of high pressure to one of lower pressure. No matter how much the exhaust pressure is reduced, the velocity of the escaping gas cannot exceed its sonic velocity.

It is possible to have the rate of mass transfer at a liquid or solid surface exceed its value at the critical point. Experimental determinations of the absolute rate of vaporization are examples of such cases. This can be accomplished, however, only by removing the emitted molecules (by total condensation) while the gas is still in a nonuniform state. In this situation the problem is really no longer one of *interphase* mass transfer. Hickman (34) has defined molecular distillation as having two possible subdivisions. One is unobstructed free-path distillation, which is equivalent to evaporation of the pure substance under the conditions which have been prescribed for determining the absolute rate of vaporization. In practice, however, the second type of molecular distillation is usually encountered. This is equivalent to evaporation under such conditions that the mean free path of gas molecules is comparable to the distance between evaporating and condensing surfaces.

As the mean free path becomes smaller with respect to this distance, ordinary, or "equilibrant," distillation or evaporation is approached. It is important to recognize, however, that there is a range of operation which is neither molecular nor equilibrant in character. Departures from thermodynamic equilibrium at the phase interface are not negligible in this range.

A sharp distinction can easily be made between molecular distillation or evaporation and interphase distillation or evaporation. A true interphase process is one in which the gas phase ultimately reduces to a uniform gas, neglecting any intraphase process such as diffusion which may be occurring. As has been mentioned before, this study is concerned only with interphase mass transfer in this sense and not with strictly molecular processes.

**5.8. Comparison of the Various Theories for Interphase Mass Transfer of Monatomic Molecules.** Rates of interphase mass transfer and the corresponding density and temperature ratios for monatomic molecules have been calculated in Appendix E on the basis of equations developed in Section 5.6. Values have been calculated somewhat beyond the critical point to indicate that the rate of mass transfer would apparently decrease if that point could be passed. These results are shown graphically in figures 8 and 9. For small values of $w/w_{s+}$, the approximate equation

$$\frac{\gamma_1}{\tilde{\gamma}_s} = 1 - \frac{2}{\pi}\frac{w}{w_{s+}} \qquad (5.8\text{-}1)$$

will ordinarily be found adequate. It follows immediately from (5.6-9), since the $\phi_1^2$ terms become small quantities of the second order as $w/w_{s+} \to 0$.

Also in figures 8 and 9 is the interfacial density ratio, $\gamma_0/\tilde{\gamma}_s$, which is (with the exception of the region near the critical point in condensation processes) closer to unity than the corresponding value of the

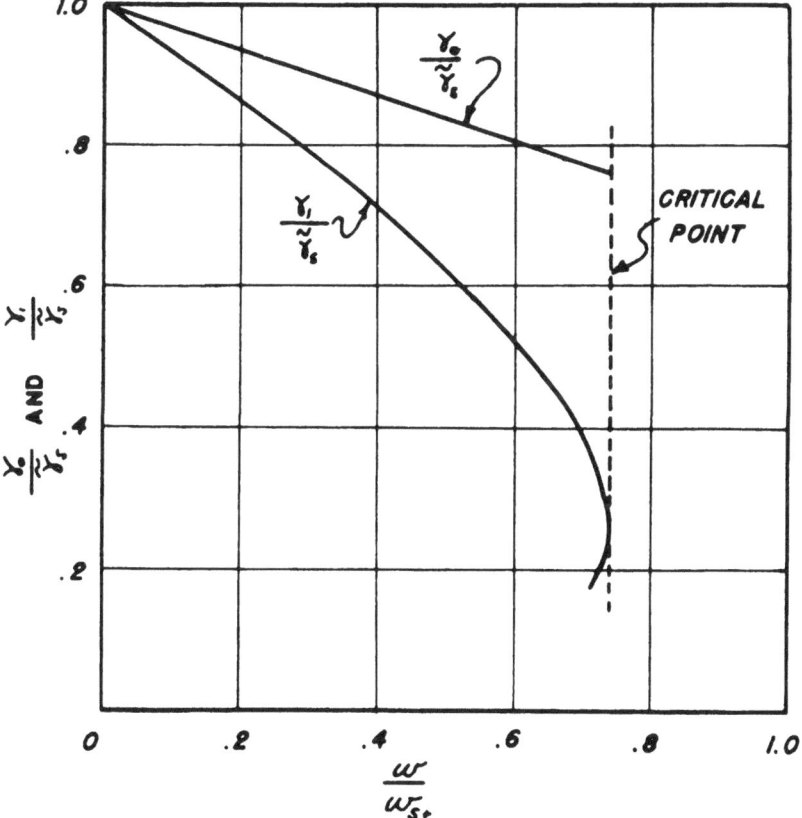

FIGURE 8.

Interfacial nonequilibrium for monatomic molecules in evaporation processes. Values of $w/w_{s+}$, $\gamma_0/\tilde{\gamma}_s$, and $\gamma_1/\tilde{\gamma}_s$ calculated according to the theory of Chapter V

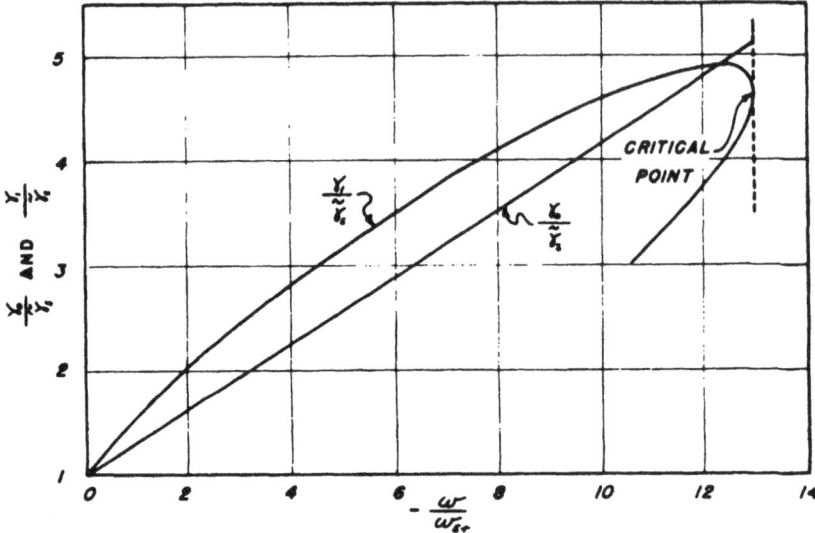

FIGURE 9.

Interfacial nonequilibrium for monatomic molecules in condensation processes. Values of $-w/w_{s+}$, $\gamma_0/\tilde{\gamma}_s$, and $\gamma_1/\tilde{\gamma}_s$ calculated according to the theory of Chapter V

density ratio at the point where the gas becomes uniform. For evaporation processes, the difference between the two ratios increases with $w/w_{s+}$. The simple theory of Chapter III makes no distinction between these two density ratios, but merely refers indefinitely to the state of the gas near the interface. It is interesting to observe that the simple theory incorporating $\Gamma$ gives results intermediate between the values of $\gamma_1/\tilde{\gamma}_s$ and $\gamma_0/\tilde{\gamma}_s$ predicted by the more elaborate theory of Chapter V. Both theories differ in numerical value rather than in order of magnitude. This is significant when the very different approaches to the problem which they represent are considered. If the factor $\Gamma$ is omitted from the simple theory, it is in greater disagreement with the theory of Chapter V.

The temperature ratio, $T_1/T_s$, according to the theory of Chapter V, is shown in figures 10 and 11. As $w/w_{s+}$ approaches zero, $T_1/T_s$ approaches unity more rapidly than the density ratios but deviates noticeably from unity as $w/w_{s+}$ is increased. It is curious to note that $T_1/T_s$ is always less than unity, regardless of the direction of the mass transfer process (evaporation or condensation). While this result was not expected, there is no particular reason to conclude that it is unreasonable.

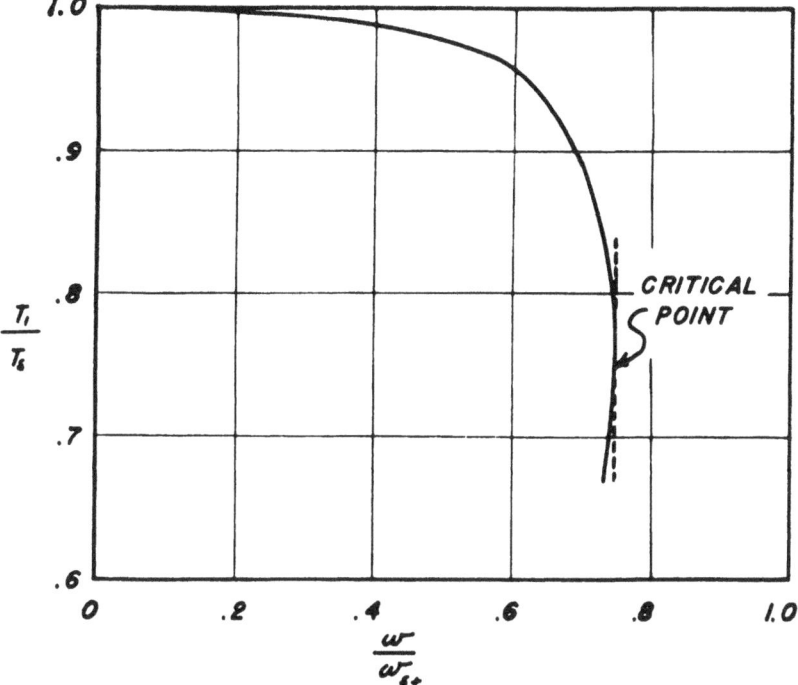

FIGURE 10.

Interfacial nonequilibrium for monatomic molecules in evaporation processes. Values of $w/w_{s+}$ and $T_1/T_0$ calculated according to the theory of Chapter V

**5.9. Extension of the Theory to Polyatomic Molecules.** From the viewpoint of the present discussion, the major difference between a monatomic and polyatomic molecule is that the latter may possess forms of energy other than energy of translation. In the rigorous theory of nonuniform gases, the detailed structure of polyatomic molecules affects the entire interpretation of the collision processes between molecules. However, since we have not examined these processes in the case of monatomic gases, we will not consider them for the more complex case. The extension of the theory for monatomic molecules to polyatomic molecules will, therefore, be still more approximate from a theoretical viewpoint.

If the nature of the collisions between molecules is not considered, only the energy transport equation is different for polyatomic gases. In a uniform gas the average energy of a polyatomic molecule with translatory velocity C was given by (1.2-11), which is

## 68 ANOTHER THEORY OF INTERPHASE MASS TRANSFER

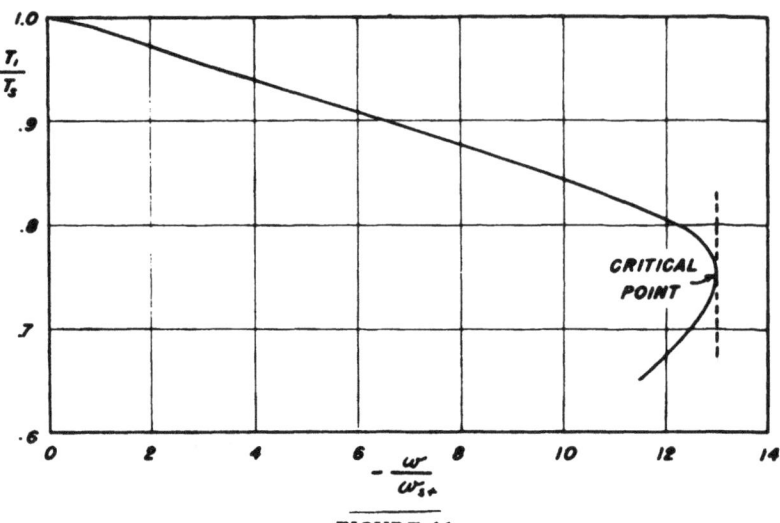

FIGURE 11.

Interfacial nonequilibrium for monatomic molecules in condensation processes. Values of $-w/w_{s+}$ and $T_1/T_s$ calculated according to the theory of Chapter V

$$\epsilon = \frac{1}{2}mC^2 + \frac{m}{4\beta^2}(s - 3), \tag{5.9-1}$$

where $s$ is defined as in Section 1.2b. The second term of (5.9-1) is not a function of $C$, so that for the uniform gas the rate of energy transport may be calculated as

$$E_{tot} = \int_{-\infty}^{\infty}\int_{-\infty}^{\infty}\int_{-\infty}^{\infty} U\left[\frac{1}{2}mC^2 + \frac{m}{4\beta_1^2}(s - 3)\right] f_1 dUdVdW. \tag{5.9-2}$$

This differs from (5.5-8) only in the additional term due to the internal energy of the molecules:

$$E_i = \int_{-\infty}^{\infty}\int_{-\infty}^{\infty}\int_{-\infty}^{\infty} \frac{Um}{4\beta_1^2}(s - 3)f_1 dUdVdW. \tag{5.9-3}$$

Since the group denoting the internal energy of a molecule is constant, (5.9-3) is simply

## ANOTHER THEORY OF INTERPHASE MASS TRANSFER

$$E_i = \frac{n_1 m U_1}{4\beta_1^2}(s-3),  \quad (5.9\text{-}4)$$

and from (5.5-9) and (5.9-4) we obtain the rate of energy transport in the uniform polyatomic gas:

$$E_{tot} = \frac{\gamma_1 U_1}{2\beta_1^2}\left[\frac{5}{2} + \beta_1^2 U_1^2 + \frac{s-3}{2}\right]. \quad (5.9\text{-}5)$$

According to the theory for the absolute rate of vaporization (Chapter II), the velocity and energy distribution of those molecules which move from the phase interface is the same as in a uniform gas. Thus, (5.9-1) may be used with the distribution function of these molecules to calculate the energy flow from the phase interface. It will be assumed that (5.9-1) also gives the energy of the molecules moving toward the phase interface. This is probably approximately true. Then the rate of energy transport at the phase interface is, from (5.4-2a),

$$E_{tot} = \int_{-\infty}^{\infty}\int_{-\infty}^{\infty}\int_{-\infty}^{\infty} U\left[\frac{1}{2}mC^2 + \frac{m}{4\beta_s^2}(s-3)\right]\tilde{f}_s dU dV dW$$

$$+ B\int_{-\infty}^{\infty}\int_{-\infty}^{\infty}\int_0^{\infty} U^2\left[\frac{1}{2}mC^2 + \frac{m}{4\beta_s^2}(s-3)\right]\tilde{f}_s dU dV dW. \quad (5.9\text{-}6)$$

This differs from (5.4-16) only in the additional term due to internal energy:

$$E_i = \int_{-\infty}^{\infty}\int_{-\infty}^{\infty}\int_{-\infty}^{\infty} U\left[\frac{m}{4\beta_s^2}(s-3)\right]\tilde{f}_s dU dV dW$$

$$+ B\int_{-\infty}^{\infty}\int_{-\infty}^{\infty}\int_0^{\infty} U^2\left[\frac{m}{4\beta_s^2}(s-3)\right]\tilde{f}_s dU dV dW, \quad (5.9\text{-}7)$$

and, since the group $[(m/4\beta_s^2)(s-3)]$ is constant, we have

$$E_i = \left[\frac{m}{4\beta_s^2}(s-3)\right]\int_{-\infty}^{\infty}\int_{-\infty}^{\infty}\int_{-\infty}^{\infty} U\tilde{f}_s dU dV dW$$

$$+ \left[\frac{m}{4\beta_s^2}(s-3)\right]B\int_{-\infty}^{\infty}\int_{-\infty}^{\infty}\int_0^{\infty} U^2\tilde{f}_s dU dV dW. \quad (5.9\text{-}8)$$

## 70 ANOTHER THEORY OF INTERPHASE MASS TRANSFER

The first integral of (5.9-8) is zero, since there is no net mass transfer in the equilibrium gas represented by $f_s$. The second integral was evaluated in (5.4-5), and, from the result obtained there, we get

$$E_i = \frac{\tilde{\gamma}_s}{16\beta_s^4}(s-3)B, \qquad (5.9\text{-}9)$$

and, when $B$ is substituted from (5.4-8), this becomes

$$E_i = \frac{\tilde{\gamma}_s}{8\beta_s^3}(s-3)\frac{w}{w_{s+}}, \qquad (5.9\text{-}10)$$

which can be combined with (5.4-26) to give the total rate of energy transport as

$$E_{tot} = \frac{\tilde{\gamma}_s}{8\beta_s^3}\frac{w}{w_{s+}}(s+2). \qquad (5.9\text{-}11)$$

Equating (5.9-5) and (5.9-11) gives

$$\frac{\gamma_1 \beta_1 U_1}{2\beta_1^3}\left[\frac{s+2}{2} + \beta_1^2 U_1^2\right] = \frac{w}{w_{s+}}(s+2)\frac{\tilde{\gamma}_s}{8\beta_s^3}, \qquad (5.9\text{-}12)$$

$$\left(\frac{\gamma_1}{\tilde{\gamma}_s}\right)\left(\frac{T_1}{T_s}\right)^{3/2} 4\pi^{1/2}\beta_1 U_1\left[\frac{s+2}{2} + \beta_1^2 U_1^2\right] = \frac{w}{w_{s+}}(s+2), \qquad (5.9\text{-}13)$$

and expressing $\beta_1 U_1$ as $\phi_1$ from (5.6-2) gives

$$\left(\frac{\gamma_1}{\tilde{\gamma}_s}\right)\left(\frac{T_1}{T_s}\right)^{3/2} 4\pi^{1/2}\frac{1}{2\pi^{1/2}}\frac{w}{w_{s+}}\left(\frac{T_1}{T_s}\right)^{-1/2}\left(\frac{\gamma_1}{\tilde{\gamma}_s}\right)^{-1}\left[\frac{s+2}{2} + \phi_1^2\right]$$

$$= \frac{w}{w_{s+}}(s+2), \qquad (5.9\text{-}14)$$

$$\left(\frac{T_1}{T_s}\right)\left[\frac{s+2}{2} + \phi_1^2\right] = \frac{s+2}{2}, \qquad (5.9\text{-}15)$$

## ANOTHER THEORY OF INTERPHASE MASS TRANSFER

$$\frac{T_1}{T_s} = \left[1 + \frac{2\phi_1^2}{s+2}\right]^{-1}, \qquad (5.9\text{-}16)$$

which is the equation for polyatomic gases equivalent to (5.6-8). Substituting (5.9-16) in (5.6-3), we obtain

$$\frac{\gamma_1}{\hat{\gamma}_s} = \left[1 - \frac{2}{\pi}\frac{w}{w_{s+}}\right]\frac{1 + 2\phi_1^2/(s+2)}{1 + 2\phi_1^2}. \qquad (5.9\text{-}17)$$

For small values of $w/w_{s+}$ the $\phi_1^2$ terms in (5.9-17) are negligible. In this case there is no difference in the departure from equilibrium density for the monatomic and polyatomic gas.

## CHAPTER VI

## PREDICTED DEPARTURES FROM INTERFACIAL EQUILIBRIUM IN TYPICAL MASS TRANSFER PROCESSES

**6.1. Interphase Mass Transfer at Ordinary Pressures.** Experimental work which was intended to examine interphase mass transfer directly has been discussed in earlier parts of this study. With these exceptions, most of the investigations of "interphase" mass transfer have really been studies of intraphase transfer, the interpretation of which is dependent upon the familiar assumption of negligible departure from thermodynamic equilibrium at phase interfaces. While on the whole a certain amount of success has resulted from this viewpoint, all results have not been conclusive enough to leave no doubt as to the validity of this assumption.

In this and the following section, part of the theory previously developed is applied to two sets of typical interphase mass transfer data. The first of these, to be discussed in this section, involves a gas phase at nearly atmospheric pressure. The second involves a gas phase at considerably lower pressure. None of this work permits any confirmation of the theory of this study. The only purpose in discussing it here is to illustrate the application of the theory to real cases and to show the order of magnitude of the effects which it predicts.

Since the rate of mass transfer per unit of interfacial area is an important factor in determining nonequilibrium at phase interfaces, it is clear that the only situations which can be analyzed from this viewpoint are those where the interfacial area is known. One of the most convenient methods for obtaining such data on mass transfer between liquids and gases is the wetted-wall column. The early data of Gilliland and Sherwood (30) seem typical, so far as the range of operating conditions is concerned, of most of the later work in this sort of apparatus. Therefore, the predicted departure from interfacial equilibrium for these experiments will be examined here.

The Gilliland experiments involve the evaporation of nine nominally pure liquids into a turbulent air stream. It is better, however, to consider these liquids as saturated with air, which is probably a more accurate view of their real condition. Then the absolute rates of vaporization and condensation of air at their surfaces are equal, so that there is net mass transfer only of the second substance.

Calculations were made for one run from each of the sets of data for the nine liquids. This was chosen as the run with the highest rate of interphase mass transfer. It should not be thought, however, that departures from interfacial equilibrium are necessarily a maximum for these cases. The ratio of the actual rate of mass transfer to the absolute rate of vaporization would be a better criterion. Even this does not necessarily indicate the significance of the results, because the importance of the interfacial departure from equilibrium will usually depend upon its value relative to intraphase driving forces, rather than upon its absolute value.

In each case the calculations were based on an average temperature and an average rate of interphase mass transfer. Since the temperature varied over only a short interval, the absolute rate of vaporization does not vary much over the length of the column. The rate of mass transfer is subject to greater variation, but, since only the magnitude of the effect is of interest here, basing the calculations on the average rate is not of importance.

The details of the calculation are tabulated in Appendix F, and the results are summarized in Table II. The predicted values of

TABLE II

Interfacial nonequilibrium for the experiments of Gilliland and Sherwood (30). Based on runs at highest rate of interphase mass transfer.

| Liquid | Run No. | $P_0$ mm Hg | $T_s$ °R | $\omega_A$ lb-mols hr-ft$^2$ | $y_{A_s}^* - y_{A_0}$ $\times 10^5$ | $y_{A_s}^* - y_{A_0}$ as percent of total driving force |
|---|---|---|---|---|---|---|
| Water | P75p | 574 | 585 | 0.1715 | 4.50 | 0.044 |
| s-Butanol | Sa5p | 782 | 552 | 0.0288 | 1.23 | 0.033 |
| n-Butanol | NB27c | 396 | 582 | 0.0340 | 2.79 | 0.055 |
| Toluene | T19c | 767 | 582 | 0.0652 | 2.91 | 0.035 |
| Aniline | A21p | 770 | 688 | 0.0530 | 2.70 | 0.045 |
| Chlorobenzene | C3p | 795 | 585 | 0.0349 | 1.82 | 0.047 |
| s-Pentanol | S5p | 800 | 573 | 0.0302 | 1.40 | 0.044 |
| Ethyl acetate | E5p | 772 | 549 | 0.0684 | 2.69 | 0.024 |
| i-Propanol | I5p | 765 | 556 | 0.0694 | 2.51 | 0.036 |

74  DEPARTURES FROM INTERFACIAL EQUILIBRIUM

$$y_{A_s}^* - y_{A_0}$$

are based on (4.2-4b). It will be noted that the predicted departure from equilibrium is very small. The driving force for mass transfer across the phase interface is only a few hundredths of a percent of the total driving force in these experiments. These calculations are based on $\sigma_A = 1$. Even if the condensation coefficient is somewhat less than unity, no significant value of

$$y_{A_s}^* - y_{A_0}$$

would be obtained. It would thus seem that, in experiments similar to these, substantially complete equilibrium exists at phase interfaces.

**6.2. Interphase Mass Transfer at Reduced Pressures.** Interphase mass transfer at reduced pressures has long been an important part of commercial processes. There evidently have not been many quantitative experimental examinations of mass transfer in this low pressure range. Several studies have been made, for example, that of Berg and Popovac (12), of the efficiency of rectifying columns at low pressures, but these results do not permit any analysis in terms of specific rates of mass transfer per unit of interfacial area. Considerable disagreement has been voiced regarding even the qualitative behavior of columns at reduced pressure.[1] It might be mentioned here that equilibrium data for low pressure systems have been typically secured in "dynamic" equilibrium stills. It would be of considerable interest to see if such "equilibrium" data were a function of the rate of distillation in the "equilibrium" still.

A recent paper by Tucker and Sherwood (91) on vacuum dehydration using liquid absorbents contains data suitable for quantitative discussion because interphase mass transfer rates are known. A wetted-wall column was used to remove water from a vacuum drying system by absorption in 51 percent lithium bromide solution at approximately 10°C. The rate of water absorption was said to be controlled by diffusion through the liquid film. "At the low pressures attained, the rate of diffusion in the gas phase is extremely rapid, and the vapor and liquid may be assumed to be in equilibrium, even when several percent of air is present." It will be shown below, however, that, even if there is no gas "film" resistance, the resistance to mass transfer across the phase interface may itself be appreciable.

[1] See discussion following reference 12.

The data which are actually reported in the paper are for cases in which there is only water in the gas phase. Since the lithium bromide solution consists only of water and the nonvolatile salt, the data can conveniently be analyzed in terms of the theory for interphase mass transfer of a pure substance. From this viewpoint, the only effect of the lithium bromide on the system is to lower the equilibrium vapor pressure at the surface (and, of course, it affects diffusional transport of water in the solution). The data have been calculated in Appendix G on the basis of the theory of Chapter III. An interpretation might also be given in terms of Chapter V, but, since only the magnitude of these effects are of interest here, the simpler analysis of Chapter III will be used.

The results are summarized in Table III. It will be noted that, for the four cases for which data are available, the departure from equilibrium at the phase interface ranges from 0.8 to 3.0 percent of the total

TABLE III

Interfacial equilibrium for the experiments of Tucker and Sherwood(91).

| $-w$ $\frac{lb}{hr\text{-}ft^2}$ | $P_0$ mm Hg | $P_0 - P^*_{bulk}$ mm Hg | $P_0 - P^*_s$ mm Hg | Interfacial nonequilibrium as percent of total nonequilibrium |
|---|---|---|---|---|
| 2.75  | 6.1 | 1.6 | 0.0129  | 0.8 |
| 1.94  | 4.2 | 0.5 | 0.00905 | 1.8 |
| 1.29  | 3.2 | 0.4 | 0.00600 | 1.5 |
| 0.645 | 2.1 | 0.1 | 0.00298 | 3.0 |

driving force for transfer between the liquid and gas phases. These figures are to be regarded as approximate, since they are based on data which were read from graphs.

Although, if interfacial nonequilibrium were only of this order of magnitude for these experiments, it might well be neglected, it must be remembered that the actual departure from equilibrium may be considerably greater if the condensation coefficient is not equal to unity. In fact, if $\sigma = 0.04$ for water, as Alty and Prüger have claimed (although it is not believed that this has been conclusively demonstrated), then the driving force across the phase interface may amount to all of the driving force noted by Tucker and Sherwood.[2]

The range of operating conditions encountered in these experiments is becoming of increasing industrial importance. The low temperature vacuum drying processes used for dehydration of foodstuffs and certain pharmaceutical products are the outstanding examples at the moment. Carman (20) has emphasized the importance of an adequate theoretical interpretation of interphase mass transfer processes in this range of operation; his own analysis is based on the simple considerations which led to equation (3.2-4). Others (17) have occasionally analyzed limited experimental data in terms of the theory of the absolute rate of vaporization and consequently obtained very low values for the condensation coefficient (see footnote 5 of Chapter II).

For the most part, the experimental investigations of mass transfer under conditions where the theory of true interphase mass transfer is important have been made from the viewpoint of improving the technical "art" of industrial processes. It is hoped that there will be future experimental work along more fundamental lines and that the theory of this study will find application in its interpretation.

[2]Since some diffusional resistance is to be expected in the liquid phase, this may indicate that the value of $\sigma$ cannot be as low as reported by Alty and Prüger.

## CHAPTER VII

### CONCLUSION

In conclusion, it would be well to summarize the most important points developed in this study. The review given here will serve both to emphasize the relation of the several parts of this investigation to each other and to distinguish between this and earlier contributions to the literature of interphase mass transfer theory.

Chapter I was devoted to an account of the early history of the problem and a discussion of certain ideas which are important to the later theoretical development. From these preliminary concepts, the theory of the absolute rate of vaporization was formulated in Chapter II. Although the central idea of this treatment has been used extensively before to predict the maximum rate of vaporization of liquids and solids, the theory of the subject has not been examined in such detail and some of the refinements incorporated into the discussion are new.

In Chapter III a relatively simple theory for interphase mass transfer of a pure substance was developed. It was shown that the gas temperature during an evaporation or condensation process is not necessarily identical with the temperature of the liquid or solid surface. This fact has not been generally recognized and leads to a form of equation for the interphase mass transfer rate which differs from that used in many other works. These other treatments also fail to allow for the mass motion of gas relative to the phase interface which must occur in interphase mass transfer processes. Consideration of this effect led to the use of a correction factor which had been used in a special context in a paper by Slepian and Brubaker (85). The corrected simple theory was finally expressed in terms of a dimensionless equation which can be conveniently graphed or tabulated, provided the interphase temperature difference is known or assumed.

The simple theory for a pure substance was next extended to multicomponent systems, and special attention was devoted to binary mixtures. This treatment also allows for the interfacial temperature difference and for mass motion of gas to or from the phase interface. The departure of the mol fraction of a component in the gas phase from its equilibrium value does not depend upon the interfacial temperature or density departure from equilibrium. This fact is of importance, since often these latter quantities are of lesser direct interest and cannot be predicted entirely satisfactorily. The theory of Chapter IV is almost entirely new, since the only previous attempts to discuss multicomponent systems have resulted in equations for each component which are identical in form to the uncorrected simple theory for the pure substance.

In Chapter V the special case of interphase mass transfer of a pure substance with no heat transfer in the gas was considered in more detail and the most objectionable approximation of Chapter III eliminated. An empirical correction factor is applied to the equilibrium velocity distribution of the gas at the phase interface and permits the behavior of the gas to be described rigorously in terms of its distribution function. This method of analysis was suggested in part by the work of Crout (23) and of Giddings and Crout (29). However, the form of distribution function which they chose to use was shown in Section 5.2 to be unreasonable, and hence the details of the two treatments are entirely different.

Crout has shown that there is a maximum rate of evaporation which cannot be exceeded if the evaporating gas becomes uniform. This argument was discussed in Section 5.7 and was also applied to condensation. It was used in this study to make a distinction between the process of true interphase mass transfer and molecular evaporation or condensation. This distinction is independent of any assumption as to the behavior of the gas at the phase interface, but the rate of interphase mass transfer at this critical point does depend on the exact description of the gas at the interface.

The theory of Chapter V was initially developed for monatomic molecules and later extended more approximately to polyatomic molecules. The distinction between the two cases is important only when the rate of mass transfer is comparable in magnitude with the absolute rate of vaporization. The theory developed in Chapter V was expressed in terms of dimensionless equations, and these were graphed and tabulated for the case of monatomic molecules. These results cannot be compared directly with those of Chapter III, because this earlier treatment did not distinguish between the condition of the gas just at the phase interface and at the point where the interface no longer exerts any influence on the behavior of the gas. The fully corrected simple theory, however, with the gas and liquid or solid temperatures assumed equal, indicates an equilibrium departure for the gas density which is generally between the two values predicted by the more elaborate theory. The uncorrected simple theory predicts larger equilibrium departures than either of the other theories. It is felt, therefore, that in a sense the approximate agreement between the fully corrected simple theory and the theory of Chapter V lends a certain amount of support to the adequacy of the former treatment for many situations and indicates its superiority to the uncorrected simple theory.

# CONCLUSION

Since the theory for multicomponent systems as presented in Chapter IV is very similar in its assumptions to the corrected simple theory for the pure substance, it would seem that this (multicomponent) theory is also somewhat supported by the approximate agreement between Chapters III and V. It might be thought that it would be desirable to extend the theory of Chapter V to multicomponent mixtures. This, however, is not possible without much more involved considerations. It will be recalled that steady-state conditions permitted three equations to be deduced from the three summational invariants of a monatomic gas. From these three equations, the state of the uniform gas away from the phase interface (fixed by three variables which may be density, temperature, and mean velocity) was calculated. In the case of a multicomponent gas, however, another parameter, namely, composition, is involved. Furthermore, the gas does not in general reduce to the uniform state as it moves from the interface, but instead to the much more complicated condition in which diffusion is occurring. An interpretation of binary mixtures similar to the treatment of Chapter V would necessarily involve a consideration of the diffusional mechanism occurring in the gas in the region between the phase interface and the point where the interface has no direct influence on the behavior of the gas.

Direct experimental investigations of interphase mass transfer have not provided any very useful information, evidently, largely because of the lack of knowledge as to the nature of the phenomenon under examination. For the pure substance, it is important to measure (1) the rate of interphase mass transfer, (2) the temperature of the liquid or solid surface, and (3) the temperature and pressure of the gas. Provided substantially all the energy necessary for evaporation or condensation is transferred through the condensed phase, it is not necessary to measure the gas properties precisely at the phase interface. The uniform gas removed from the interface by several mean free paths is not expected to vary with distance. Its properties would be a logical first check on the theory of interphase mass transfer of the pure substance. It seems hardly necessary to emphasize the advisability of conducting the investigation under conditions where the predicted departure from equilibrium is large enough to be measurable. The steady-state apparatus of Prüger (71), perhaps with some modifications, seems well suited to further experimental work along these lines.

The difficulty in securing point analyses of composition apparently precludes the use of similar experimental methods for binary systems. The most useful work in this case might be done by methods similar to those ordinarily used in chemical engineering mass transfer studies (for example, the work described in Chapter VI). The operating conditions would, of course, be chosen so that interfacial effects would be expected to be appreciable, and then the data obtained could be analyzed on the basis both of assumed interfacial equilibrium and of the departure theoretically predicted. A comparison of the two interpretations may indicate which is a more correct working hypothesis. Unfortunately, however, many extraneous considerations complicate this type of study.

## LIST OF SYMBOLS

(Symbols used in a limited context are not included in the list below.)

*English Letters*

- $B$ constant defined by (5.4-8)
- $C$ molecular velocity
- $E$ rate of energy flow (energy/unit time-unit area)
- $f$ velocity distribution function defined by (1.2-1)
- $\mathbf{i}, \mathbf{j}, \mathbf{k}$ unit vectors in the $x, y, z$ directions
- $m$ mass of a molecule
- $M$ molecular weight
- $n$ number density of molecules
- $\mathbf{n}$ unit vector
- $P$ pressure
- $\mathbb{P}$ pressure tensor defined by (1.2-16)
- $\mathbb{P}_s$ surface pressure tensor defined by (1.2-18)
- $\mathbf{r}$ radius vector
- $R$ perfect gas constant
- $s$ number of squared terms in the expression for the total energy of a gas molecule
- $T$ absolute temperature
- $U, V, W$ components of the molecular velocity $C$ in the $x, y, z$ directions
- $w$ rate of interphase mass transfer (mass/unit time-unit area)
- $\sigma w_{s+}$ absolute rate of vaporization (mass/unit time-unit area)
- $x, y, z$ Cartesian coordinate axes
- $y_A$ mol fraction of the component $A$

*Greek Letters*

- $\alpha$ thermal accommodation coefficient defined by (1.5-6)
- $\beta$ dimensional constant defined by (1.2-8)

# LIST OF SYMBOLS

$\gamma$    mass density

$\Gamma$    dimensionless correction factor defined by (3.1-10), or by (4.2-5) if used with subscript

$\epsilon$    energy of a molecule

$\xi$    viscous slip coefficient defined by (1.5-1)

$\nu$    specular reflection coefficient (see Section 1.4)

$\zeta$    temperature jump coefficient defined by (1.5-2)

$\rho$    molal density

$\sigma$    condensation coefficient (see Section 2.1)

$\phi$    dimensionless parameter defined by (3.1-15), (4.2-7), or (5.6-2), depending on subscript

$\psi$    any molecular property

$\omega$    molal rate of interphase mass transfer (mols/unit area-unit time)

$\sigma\omega_{s+}$    molal absolute rate of vaporization (mols/unit area-unit time)

*Subscripts*

$A, B$, etc.    the components of a multicomponent system

$s$    surface of liquid or solid

$0$    gas just at the phase interface

$1$    uniform gas some distance from the phase interface

$+$    molecules with $U > 0$

$-$    molecules with $U < 0$

Vectors are given in boldface type, and the same letter in ordinary (italic) type denotes the magnitude of the vector.

An overscore indicates that the mean of the quantity beneath it is to be taken. The tilde denotes the properties of a gas phase which would be in equilibrium with a liquid or solid surface, and an asterisk those which would be in equilibrium with a liquid or solid surface which exists under its own vapor pressure.

Any mutually consistent set of *absolute* dimensional units may be used in all equations. Readers are cautioned against using engineering units without making appropriate use of the gravitational conversion factor, $g_c$.

# APPENDIX A

## Table of Frequently Used Integrals[1]

$$\frac{2}{\pi^{1/2}} \int_0^x e^{-x^2} dx = \Phi(x) \tag{A1}$$

$$\int_0^\infty e^{-\beta^2 x^2} dx = \frac{\pi^{1/2}}{2\beta} \tag{A2}$$

$$\int_0^\infty x e^{-\beta^2 x^2} dx = \frac{1}{2\beta^2} \tag{A3}$$

$$\int_0^\infty x^2 e^{-\beta^2 x^2} dx = \frac{\pi^{1/2}}{4\beta^3} \tag{A4}$$

$$\int_0^\infty x^3 e^{-\beta^2 x^2} dx = \frac{1}{2\beta^4} \tag{A5}$$

$$\int_0^\infty x^4 e^{-\beta^2 x^2} dx = \frac{3\pi^{1/2}}{8\beta^5} \tag{A6}$$

$$\int e^{ax} dx = \frac{e^{ax}}{a} + \text{constant} \tag{A7}$$

$$\int_{-\infty}^\infty \int_{-\infty}^\infty \int_{-\infty}^\infty r^2 e^{-\beta^2 r^2} dx\,dy\,dz = \frac{3\pi^{3/2}}{2\beta^5}, \text{ where } r^2 = x^2 + y^2 + z^2 \tag{A8}$$

---

[1] The probability or error integral, $\Phi(x)$, is tabulated as a function of $x$ in reference 36.

## APPENDIX B

The correction factor $\Gamma$ was defined for a pure substance by (3.1-10) and for a component of a multicomponent gas by (4.2-5). Both of these definitions are of the form

$$\Gamma = e^{-\phi^2} - \phi\pi^{1/2}\left[1 - \Phi(\phi)\right], \qquad (B1)$$

where $\phi$ is a group given by (3.1-15) or (4.2-7) for each of the respective cases.

In Table IV, $\Gamma$ is given as a function of $\phi$ for the range $10^{-3} \leq |\phi| \leq 1$. Values of $\exp(-\phi^2)$ and $\Phi(\phi)$, the probability or error

### TABLE IV
The correction factor $\Gamma$ as a function of $\phi$.

| $|\phi|$ | $e^{-\phi^2}$ | $\pm\Phi(\pm\phi)$ | $\Gamma(+\phi)$ | $\Gamma(-\phi)$ |
|---|---|---|---|---|
| 1 | 0.3679 | 0.8427 | 0.089 | 3.634 |
| 0.8 | 0.5273 | 0.7421 | 0.162 | 2.998 |
| 0.6 | 0.6977 | 0.6039 | 0.276 | 2.403 |
| 0.4 | 0.8521 | 0.4284 | 0.447 | 1.865 |
| 0.2 | 0.9608 | 0.2227 | 0.685 | 1.394 |
| 0.1 | 0.9900 | 0.1125 | 0.833 | 1.187 |
| 0.08 | 0.9936 | 0.0901 | 0.8646 | 1.1482 |
| 0.06 | 0.9964 | 0.0676 | 0.8972 | 1.1100 |
| 0.04 | 0.9984 | 0.0451 | 0.9307 | 1.0725 |
| 0.02 | 0.9996 | 0.0226 | 0.9649 | 1.0359 |
| 0.01 | 0.9999 | 0.0113 | 0.9824 | 1.0178 |
| 0.001 | 0.999999 | 0.00113 | 0.998229 | 1.001773 |

integral, were taken from Jahnke and Emde's *Tables of Functions* (36). For values of $|\phi| < 10^{-3}$, $\Gamma$ may be estimated from the approximate equation

$$\Gamma = 1 - \phi\pi^{1/2}. \qquad (B2)$$

## APPENDIX C

For the special case of $T_0/T_s = 1$, equation (3.2-2) becomes

$$\frac{w}{w_{s+}} = \sigma \left[1 - \left(\frac{\gamma_0}{\tilde{\gamma}_s}\right)\Gamma\right]. \tag{C1}$$

It will be recalled that $\Gamma$ is a function of $\phi_0$, which involves both of the ratios in (C1). Direct use of this equation to calculate corresponding values of these ratios would, therefore, involve a trial-and-error computation. This can be avoided by expressing $w/w_{s+}$ as a function of the single parameter, $\phi_0$.

From the definition of $\phi_0$, equation (3.1-15), we have

$$\frac{\gamma_0}{\tilde{\gamma}_s} = \frac{1}{2\pi^{1/2}\phi_0} \frac{w}{w_{s+}}, \tag{C2}$$

and, from (C1),

$$\frac{\gamma_0}{\tilde{\gamma}_s} = \left[1 - \frac{w}{\sigma w_{s+}}\right]\frac{1}{\Gamma}. \tag{C3}$$

Equating (C2) and (C3) and solving for $w/w_{s+}$ gives

$$\frac{w}{w_{s+}} = \left[\frac{\Gamma}{2\pi^{1/2}\phi_0} + \frac{1}{\sigma}\right]^{-1} \tag{C4}$$

In Tables V and VI corresponding values of $w/w_{s+}$ and $\gamma_0/\tilde{\gamma}_s$ have been calculated for $\sigma = 1$. Each pair of values is at one of the values of $\phi$ tabulated in Table IV. It will be noted that $-w/w_{s+}$ considerably exceeds unity in some cases; this is to be expected, since there is no limit to the rate of condensation from the viewpoint of this simple theory (see, however, Section 5.7).

When $w/w_{s+}$ is small, corresponding to small values of the parameter $\phi_0$, then $\Gamma$ is given by (B2). Therefore, from (C3) we have (with $\sigma = 1$)

$$\frac{\gamma_0}{\tilde{\gamma}_s} = \left[1 - \frac{w}{w_{s+}}\right]\left[1 - \phi_0 \pi^{1/2}\right]^{-1}. \tag{C5}$$

## TABLE V

Interfacial nonequilibrium in evaporation processes. Calculated according to the theory of Chapter III for the special case of $T_0/T_s = 1$ and $\sigma = 1$.

| $\phi_0$ | $\dfrac{w}{w_{s+}}$ | $\dfrac{\gamma_0}{\bar{\gamma}_s}$ [a] | $\dfrac{\gamma_0}{\bar{\gamma}_s}$ [b] |
|---|---|---|---|
| 1 | 0.976 | 0.275 | 0.024 |
| 0.8 | 0.946 | 0.334 | 0.054 |
| 0.6 | 0.885 | 0.416 | 0.115 |
| 0.4 | 0.760 | 0.536 | 0.240 |
| 0.2 | 0.509 | 0.717 | 0.491 |
| 0.1 | 0.299 | 0.842 | 0.701 |
| 0.08 | 0.247 | 0.871 | 0.753 |
| 0.06 | 0.192 | 0.901 | 0.808 |
| 0.04 | 0.132 | 0.932 | 0.868 |
| 0.02 | 0.0685 | 0.965 | 0.932 |
| 0.01 | 0.0348 | 0.983 | 0.965 |
| 0.001 | 0.00354 | 0.9983 | 0.9965 |

[a] See equation (3.2-2). [b] See equation (3.2-2) with $I \equiv 1$.

Under these same circumstances $\phi_0$ is given approximately by (3.1-15a). Therefore, (C5) becomes

$$\frac{\gamma_0}{\bar{\gamma}_s} = \left[1 - \frac{w}{w_{s+}}\right]\left[1 - \frac{1}{2}\frac{w}{w_{s+}}\right]^{-1}. \tag{C6}$$

Since $w/w_{s+}$ is a small quantity, this may be written

$$\frac{\gamma_0}{\bar{\gamma}_s} = 1 - \frac{1}{2}\frac{w}{w_{s+}}. \tag{C7}$$

## TABLE VI

Interfacial nonequilibrium in condensation processes. Calculated according to the theory of Chapter III for the special case of $T_0/T_s = 1$ and $\sigma = 1$.

| $-\phi_0$ | $\dfrac{w}{w_{s+}}$ | $\dfrac{\gamma_0}{\gamma_s}$ [a] | $\dfrac{\gamma_0}{\bar{\gamma}_s}$ [b] |
|---|---|---|---|
| 1 | 39.8 | 11.2 | 40.8 |
| 0.8 | 17.5 | 6.2 | 18.5 |
| 0.6 | 7.70 | 3.62 | 8.70 |
| 0.4 | 3.17 | 2.24 | 4.17 |
| 0.2 | 1.04 | 1.460 | 2.04 |
| 0.1 | 0.426 | 1.201 | 1.426 |
| 0.08 | 0.328 | 1.157 | 1.328 |
| 0.06 | 0.237 | 1.114 | 1.237 |
| 0.04 | 0.152 | 1.075 | 1.152 |
| 0.02 | 0.0735 | 1.036 | 1.0735 |
| 0.01 | 0.0361 | 1.018 | 1.0361 |
| 0.001 | 0.00355 | 1.0018 | 1.0036 |

[a] See equation (3.2-2). [b] See equation (3.2-2) with $\Gamma \equiv 1$.

For most purposes (C7) is an adequate approximation for values of $w/w_{s+}$ smaller than those in Tables V and VI.

Also tabulated, in the last columns of Tables V and VI, are the values of $\gamma_0/\bar{\gamma}_s$ which correspond to each of the values of $w/w_{s+}$ if the considerations which led to $\Gamma$ are neglected. Thus, the equation on which these values are based is (C3) with $\sigma = 1$ and $\Gamma \equiv 1$:

$$\frac{\gamma_0}{\bar{\gamma}_s} = 1 - \frac{w}{w_{s+}}. \tag{C8}$$

This is equivalent to (3.2-4) with $\sigma = 1$. (There is, of course, no connection between values of $\gamma_0/\bar{\gamma}_s$ calculated from this equation and the values of $\phi_0$.)

## APPENDIX D

The rate of energy transport in a uniform monatomic gas with the linear velocity $U_1$ was given by (5.5-8) as

$$E = \int_{-\infty}^{\infty} \int_{-\infty}^{\infty} \int_{-\infty}^{\infty} U\left(\frac{1}{2}mC^2\right) f_1 dU dV dW. \tag{D1}$$

The quantity $UC^2$ may be written

$$UC^2 = U(U^2 + V^2 + W^2) \tag{D2}$$

or

$$UC^2 = (U - U_1)[(U - U_1)^2 + V^2 + W^2] + H, \tag{D3}$$

where $H$ is a correction term. Equating (D2) and (D3) and solving for $H$ gives

$$H = U_1[(U - U_1)^2 + V^2 + W^2] + 2U_1 U^2 - U_1^2 U. \tag{D4}$$

Then, from (D3) and (D4), equation (D1) may be written

$$E = \frac{1}{2}m \int_{-\infty}^{\infty} \int_{-\infty}^{\infty} \int_{-\infty}^{\infty} (U - U_1)[(U - U_1)^2 + V^2 + W^2] f_1 dU dV dW$$

$$+ \frac{1}{2}mU_1 \int_{-\infty}^{\infty} \int_{-\infty}^{\infty} \int_{-\infty}^{\infty} [(U - U_1)^2 + V^2 + W^2] f_1 dU dV dW$$

$$+ mU_1 \int_{-\infty}^{\infty} \int_{-\infty}^{\infty} \int_{-\infty}^{\infty} U^2 f_1 dU dV dW$$

$$- \frac{1}{2}mU_1^2 \int_{-\infty}^{\infty} \int_{-\infty}^{\infty} \int_{-\infty}^{\infty} U f_1 dU dV dW. \tag{D5}$$

The first integral of (D5) is the rate of energy transport through a plane moving with the linear velocity of a uniform gas. This is, therefore, zero. The remaining integrals may be calculated individually as follows:

$$\frac{1}{2}mU_1 \int_{-\infty}^{\infty}\int_{-\infty}^{\infty}\int_{-\infty}^{\infty}[(U-U_1)^2+V^2+W^2]f_1 dUdVdW$$

$$=\frac{1}{2}\gamma_1 U_1\left(\frac{\beta_1^3}{\pi^{3/2}}\right)\int_{-\infty}^{\infty}\int_{-\infty}^{\infty}\int_{-\infty}^{\infty}[(U-U_1)^2+V^2+W^2]$$

$$\times e^{-\beta_1^2[(U-U_1)^2+V^2+W^2]} dUdVdW. \quad (D6)$$

This integral is given by (A8), since $dU = d(U - U_1)$, so that (D6) becomes

$$=\frac{1}{2}\gamma_1 U_1\left(\frac{\beta_1^3}{\pi^{3/2}}\right)\frac{3}{2}\frac{\pi^{3/2}}{\beta_1^5} \quad (D7)$$

$$=\frac{3}{4}\frac{\gamma_1 U_1}{\beta_1^2}. \quad (D8)$$

The third integral of (D5) follows from (5.5-2) and (5.5-7):

$$mU_1\int_{-\infty}^{\infty}\int_{-\infty}^{\infty}\int_{-\infty}^{\infty}U^2 f_1 dUdVdW = \frac{\gamma_1 U_1}{2\beta_1^2}[1+2\beta_1^2 U_1^2]. \quad (D9)$$

The last integral of (D5) is simply

$$\frac{1}{2}mU_1^2\int_{-\infty}^{\infty}\int_{-\infty}^{\infty}\int_{-\infty}^{\infty}Uf_1 dUdVdW = \frac{1}{2}\gamma_1 U_1^3. \quad (D10)$$

Combining (D8), (D9), and (D10) to obtain (D5), we have

$$E = \frac{\gamma_1 U_1}{2\beta_1^2}\left[\frac{3}{2}+1+2\beta_1^2 U_1^2 - \beta_1^2 U_1^2\right] \quad (D11)$$

$$= \frac{\gamma_1 \beta_1 U_1}{2\beta_1^3}\left[\frac{5}{2}+\beta_1^2 U_1^2\right]. \quad (D12)$$

## APPENDIX E

Departures from equilibrium for monatomic molecules in evaporation and condensation processes have been calculated in Tables VII and VIII according to the theory of Chapter V. The values of $w/w_{s+}$ were

TABLE VII

Interfacial nonequilibrium for monatomic molecules in evaporation processes. Calculated according to the theory of Chapter V.

| $\phi_1$ | $\dfrac{T_1}{T_s}$ | $\dfrac{w}{w_{s+}}$ | $\dfrac{\gamma_1}{\tilde{\gamma}_s}$ | $\dfrac{\gamma_0}{\tilde{\gamma}_s}$ |
|---|---|---|---|---|
| 1.5 | 0.5263 | 0.7210 | 0.1869 | 0.7705 |
| 1.2 | 0.6345 | 0.7336 | 0.2165 | 0.7665 |
| 1.0 | 0.7143 | 0.7397 | 0.2469 | 0.7645 |
| 0.91287[a] | 0.7500 | 0.7406 | 0.2643 | 0.7643 |
| 0.9 | 0.7553 | 0.7406 | 0.2671 | 0.7643 |
| 0.8 | 0.7962 | 0.7385 | 0.2982 | 0.7649 |
| 0.7 | 0.8361 | 0.7319 | 0.3226 | 0.7670 |
| 0.6 | 0.8741 | 0.7180 | 0.3611 | 0.7714 |
| 0.5 | 0.9091 | 0.6928 | 0.4099 | 0.7795 |
| 0.4 | 0.9398 | 0.6497 | 0.4726 | 0.7932 |
| 0.3 | 0.9653 | 0.5791 | 0.5543 | 0.8157 |
| 0.2 | 0.9843 | 0.4656 | 0.6620 | 0.8518 |
| 0.1 | 0.9960 | 0.2850 | 0.8057 | 0.9093 |
| 0.05 | 0.9990 | 0.1586 | 0.8954 | 0.9495 |
| 0.01 | 0.99996 | 0.03466 | 0.9778 | 0.9890 |
| 0.005 | 0.999990 | 0.01753 | 0.9888 | 0.9944 |
| 0.001 | 0.9999996 | 0.003537 | 0.9977 | 0.9989 |

[a] Critical point.

calculated from (5.6-12), of $T_1/T_s$ from (5.6-8), of $\gamma_1/\bar{\gamma}_s$ from (5.6-11), and of $\gamma_0/\bar{\gamma}_s$ from (5.4-13). Values have been calculated somewhat beyond the critical value of $\phi_1$ to indicate that the rate of interphase mass transfer would apparently decrease if that point could be passed.

TABLE VIII

Interfacial nonequilibrium for monatomic molecules in condensation processes. Calculated according to the theory of Chapter V.

| $-\phi_1$ | $\dfrac{T_1}{T_s}$ | $-\dfrac{w}{w_{s+}}$ | $\dfrac{\gamma_1}{\bar{\gamma}_s}$ | $\dfrac{\gamma_0}{\bar{\gamma}_s}$ |
|---|---|---|---|---|
| 1.5 | 0.5263 | 8.79 | 2.28 | 3.80 |
| 1.2 | 0.6345 | 11.12 | 3.28 | 4.54 |
| 1.0 | 0.7143 | 12.72 | 4.25 | 5.05 |
| 0.91287[a] | 0.7500 | 12.98 | 4.63 | 5.13 |
| 0.9 | 0.7553 | 12.98 | 4.68 | 5.13 |
| 0.8 | 0.7962 | 12.38 | 4.89 | 4.94 |
| 0.7 | 0.8361 | 10.75 | 4.74 | 4.42 |
| 0.6 | 0.8741 | 8.372 | 4.210 | 3.665 |
| 0.5 | 0.9091 | 5.873 | 3.475 | 2.870 |
| 0.4 | 0.9398 | 3.762 | 2.736 | 2.197 |
| 0.3 | 0.9653 | 2.205 | 2.110 | 1.702 |
| 0.2 | 0.9843 | 1.143 | 1.626 | 1.364 |
| 0.1 | 0.9960 | 0.4474 | 1.265 | 1.142 |
| 0.05 | 0.9990 | 0.1988 | 1.122 | 1.063 |
| 0.01 | 0.99996 | 0.03627 | 1.023 | 1.012 |
| 0.005 | 0.999990 | 0.01793 | 1.0114 | 1.0057 |
| 0.001 | 0.9999996 | 0.003553 | 1.0023 | 1.0011 |

[a] Critical point.

# APPENDIX F

Table IX gives the principal data and results for the calculation of

$$y_{A_s}^* - y_{A_0}$$

for certain runs taken from Gilliland and Sherwood's experiments (30). These correspond to the highest rates of interphase mass transfer reported for each of the nine liquids studied. Since there was no mass transfer of air, the second component in all cases, the results are computed from (4.3-4b).

The average rate of interphase mass transfer is based on a reported interfacial area of 981 $cm^2$. The rate of mass transfer is calculated from the volumetric rate on the basis of density data at 30°C, as given by Gilliland. Since no other density data were shown, it is assumed that the volumetric rate corresponds to this temperature; if this is not the case, the error introduced is small and has no effect on the magnitude of these calculations.

The vapor pressure data used are those reported by Gilliland. With the condensation coefficient assumed equal to unity, the absolute molal rate of vaporization of component $A$ (water or the organic substance) is calculated from (4.1-3c). If $P_{A_s}^*$ is in millimeters of mercury and $T_s$ is in °R, this becomes

$$\omega_{A_{s+}} = P_{A_s}^* \frac{14.7(144)}{760} \sqrt{\frac{4.17(10^8)}{2\pi(1544)M_A T_s}} \qquad (F1)$$

$$= 576 \frac{P_{A_s}^*}{\sqrt{M_A T_s}} \frac{\text{lb-mols}}{\text{hr-ft}^2} . \qquad (F2)$$

The value of $\phi_{A_0}$ is computed from the approximate equation (4.2-7a), which is entirely satisfactory for these cases. Then $\Gamma_A$ is calculated from (B2).

The absolute rate of vaporization of air from the liquid surface is based on the "vapor pressure" of air,

$$P_{B_s}^* = P_0 - P_{A_s}^* ,$$

above the liquid, which is considered totally saturated with air at the operating pressure. The air is treated as a uniform molecular specie with a molecular weight of 29.0. Then (4.1-3c) gives

$$\omega_{B_{s+}} = P_{B_s}^* \frac{14.7(144)}{760} \sqrt{\frac{4.17(10^8)}{2\pi(1544)(29.0)T_s}} \quad \text{(F3)}$$

$$= 107.2 \frac{P_{B_s}^*}{T_s^{1/2}} \frac{\text{lb-mols}}{\text{hr-ft}^2}. \quad \text{(F4)}$$

In (F3) and (F4) $P_{B_s}^*$ is in millimeters of mercury and $T_s$ is in °R. We calculate $\phi_{B_0}$ from 4.2-7a just as we did $\phi_{A_0}$, and, similarly, obtain $\Gamma_B$ from (B2).

It will be noted that the values of $\Gamma_A - 1$, $\Gamma_B - 1$, and $\omega/\omega_{A_{s+}}$ are all small quantities, less than $10^{-3}$, for the data which have been tabulated. In this instance, it is advisable to define

$$\delta\Gamma_A = \Gamma_A - 1, \quad \text{(F5)}$$

$$\delta\Gamma_B = \Gamma_B - 1. \quad \text{(F6)}$$

Then (4.3-4b), which is given again for convenient reference as

$$\frac{y_{A_0}}{y_{A_s}^*} = \left[ y_{A_s}^* + (1 - y_{A_s}^*) \frac{\Gamma_A}{\Gamma_B(1 - \omega_A/\sigma_A \omega_{A_{s+}})} \right]^{-1}, \quad \text{(F7)}$$

may be written ($\sigma_A = 1$) as

$$\frac{y_{A_0}}{y_{A_s}^*} = \left[ y_{A_s}^* + (1 - y_{A_s}^*) \frac{1 + \delta\Gamma_A}{(1 + \delta\Gamma_B)(1 - \omega_A/\omega_{A_{s+}})} \right]^{-1}. \quad \text{(F8)}$$

Since $\delta\Gamma_A$, $\delta\Gamma_B$, and $\omega_A/\omega_{A_{s+}}$ are small quantities, this is approximately

$$\frac{y_{A_0}}{y_{A_s}^*} = \left[ y_{A_s}^* + (1 - y_{A_s}^*)\left(1 + \frac{\omega_A}{\omega_{A_{s+}}} + \delta\Gamma_A - \delta\Gamma_B\right) \right]^{-1} \quad \text{(F9)}$$

$$= 1 - (1 - y_{A_s}^*)\left(\frac{\omega_A}{\omega_{A_{s+}}} + \delta\Gamma_A - \delta\Gamma_B\right). \quad \text{(F10)}$$

## TABLE IX

Departures from interfacial equilibrium for certain runs from the experiments of Gilliland and Sherwood (30).

| Liquid | Run. No. | Air rate, g/min | Top temperature °C | Bottom temperature °C | Rate of evaporation, cc/min | Absolute pressure, mm Hg | Density at 30°C, g/cc | $\omega_A$, lb-mols / hr-ft$^2$ |
|---|---|---|---|---|---|---|---|---|
| Water | P75p | 354 | 56.2 | 46.9 | 24.8 | 574 | 0.996 | 0.1715 |
| s-Butanol | Sa5p | 330 | 34.5 | 32.7 | 21.4 | 782 | 0.797 | 0.0288 |
| n-Butanol | NB27c | 153 | 52.6 | 47.1 | 25.0 | 396 | 0.807 | 0.0340 |
| Toluene | T19c | 208 | 53.7 | 47.1 | 56.1 | 767 | 0.857 | 0.0652 |
| Aniline | A21p | 281 | 113.2 | 104.7 | 39.0 | 770 | 1.013 | 0.0530 |
| Chlorobenzene | C3p | 348 | 54.8 | 48.0 | 28.7 | 795 | 1.095 | 0.0349 |
| s-Pentanol | S5p | 371 | 49.9 | 39.6 | 26.6 | 800 | 0.802 | 0.0302 |
| Ethyl acetate | E5p | 180 | 37.0 | 25.7 | 55.3 | 772 | 0.873 | 0.0684 |
| i-Propanol | I5p | 317 | 39.2 | 31.4 | 43.0 | 765 | 0.777 | 0.0694 |

| Liquid | $T_s$ °R | $P_{A_s}^*$ mm Hg | $\omega_{A_s+}$ lb-mols / hr-ft$^2$ | $\dfrac{\omega_A}{\omega_{A_s+}} \times 10^4$ | $y_{A_s}^*$ | $\phi_{A_0} \times 10^6$ |
|---|---|---|---|---|---|---|
| Water | 585 | 100.3 | 564 | 3.04 | 0.1750 | 15.0 |
| s-Butanol | 552 | 33.2 | 94.8 | 3.04 | 0.0425 | 3.64 |
| n-Butanol | 582 | 33.5 | 92.8 | 3.66 | 0.0846 | 8.72 |
| Toluene | 582 | 95.0 | 237 | 2.75 | 0.1240 | 9.61 |
| Aniline | 688 | 67.0 | 152.9 | 3.47 | 0.0870 | 8.50 |
| Chlorobenzene | 585 | 44.9 | 101.1 | 3.45 | 0.0565 | 5.49 |
| s-Pentanol | 573 | 35.0 | 89.5 | 3.37 | 0.0437 | 4.15 |
| Ethyl acetate | 549 | 133.1 | 349 | 1.96 | 0.1725 | 9.52 |
| i-Propanol | 556 | 77.5 | 245 | 2.83 | 0.1013 | 8.07 |

TABLE IX

| Liquid | $-\delta\Gamma_A$ $\times 10^6$ | $P_{B_s}^*$ mm Hg | $\omega_{B_s+}$ $\frac{\text{lb-mols}}{\text{hr-ft}^2}$ | $\frac{\omega}{\omega_{B_s+}} \times 10^6$ | $y_{B_s}^*$ | $\phi_{B_0} \times 10^6$ | $-\delta\Gamma_B \times 10^6$ |
|---|---|---|---|---|---|---|---|
| Water | 26.6 | 474 | 2100 | 81.5 | 0.825 | 18.9 | 33.6 |
| s-Butanol | 6.45 | 749 | 3410 | 8.45 | 0.957 | 2.28 | 4.04 |
| n-Butanol | 15.5 | 362 | 1600 | 21.2 | 0.915 | 5.45 | 9.65 |
| Toluene | 17.0 | 672 | 2980 | 21.9 | 0.876 | 5.40 | 9.56 |
| Aniline | 15.1 | 703 | 2870 | 18.5 | 0.913 | 4.75 | 8.43 |
| Chlorobenzene | 9.74 | 750 | 3320 | 10.5 | 0.943 | 2.80 | 4.96 |
| s-Pentanol | 7.36 | 765 | 3410 | 8.85 | 0.956 | 2.38 | 4.21 |
| Ethyl acetate | 16.9 | 639 | 2920 | 23.4 | 0.827 | 5.45 | 9.65 |
| i-Propanol | 14.3 | 687 | 3120 | 22.2 | 0.899 | 5.62 | 9.96 |

| Liquid | $y_{A_s}^* - y_{A_0}$ $\times 10^5$ | $\Delta y_{tot}$ (top) | $\Delta y_{tot}$ (bottom) | $\Delta y$ log mean | $y_{A_s}^* - y_{A_0}$ as percent of $\Delta y$ log mean |
|---|---|---|---|---|---|
| Water | 4.50 | 0.217 | 0.0373 | 0.1022 | 0.044 |
| s-Butanol | 1.23 | 0.0446 | 0.0307 | 0.0372 | 0.033 |
| n-Butanol | 2.79 | 0.0992 | 0.0219 | 0.0511 | 0.055 |
| Toluene | 2.91 | 0.1320 | 0.0473 | 0.0825 | 0.035 |
| Aniline | 2.70 | 0.1043 | 0.0299 | 0.0595 | 0.045 |
| Chlorobenzene | 1.82 | 0.0556 | 0.0254 | 0.0386 | 0.047 |
| s-Pentanol | 1.40 | 0.0558 | 0.0159 | 0.0320 | 0.044 |
| Ethyl acetate | 2.69 | 0.232 | 0.0453 | 0.1143 | 0.024 |
| i-Propanol | 2.51 | 0.1262 | 0.0333 | 0.0697 | 0.036 |

## APPENDIX G

The data of Tucker and Sherwood (91) are given in the first few columns of Table X. The rate of interphase mass transfer in grams/minute is converted to pounds/hour-square foot by a factor of 0.43 given by the authors. The temperature of the lithium bromide solution leaving the column, the vapor pressure corresponding to its bulk composition, and the values of the gas phase pressure were read from graphs and may be subject to slight errors of interpolation.

If the condensation coefficient is considered as unity, the absolute rate of vaporization of water from the solution is calculated from (2.1-6c) as

$$w_{s+} = P_s^* \frac{14.7(144)}{760} \sqrt{\frac{4.17(10^8)18.02}{2\pi(1544)(1.8)T_s}} \tag{G1}$$

$$= 1822 \frac{P_s^*}{\sqrt{T_s}} \frac{\text{lb}}{\text{hr-ft}^2}, \tag{G2}$$

where $P_s^*$ is the surface equilibrium vapor pressure in millimeters of mercury and $T_s$ is the surface temperature in °K. These runs were made nearly isothermally, and, although it may not be entirely accurate, they will be analyzed on this basis.

For the values of $\omega/\omega_{s+}$ encountered in this work, (C7) is an adequate representation of (3.2-2). This, of course, considers that $\sigma = 1$ and $T_0/T_s = 1$. Since the system is isothermal, this may also be written (neglecting the difference between $\tilde{P}_s$ and $P_s^*$) as

$$\frac{P_0}{P_s^*} = 1 - \frac{1}{2} \frac{w}{w_{s+}}. \tag{G3}$$

Then the value of $w_{s+}$ obtained from (G2) may be substituted in (G3) to give

$$\frac{P_0}{P_s^*} = 1 - \frac{1}{2} \frac{w\sqrt{T_s}}{1822 P_s^*} \tag{G4}$$

or

$$P_0 - P_s^* = -\frac{1}{2}\frac{w\sqrt{T_s}}{1822} \text{ mm Hg} . \tag{G5}$$

The tabulated values of $P_0 - P_s^*$ were obtained from (G5).

## TABLE X

Interfacial nonequilibrium for the experiments of Tucker and Sherwood (91).

| $-w$ | | $P_0$ mm Hg | $P^*_{bulk}$ mm Hg | $P_0 - P^*_{bulk}$ mm Hg | $T_s$ °K | $P_0 - P^*_s$ mm Hg | $P_0 - P^*_s$ as percent of $P_0 - P^*_{bulk}$ |
|---|---|---|---|---|---|---|---|
| grams/min | lb/hr·ft² | | | | | | |
| 6.4 | 2.75 | 6.1 | 4.5 | 1.6 | 294 | 0.0129 | 0.8 |
| 4.5 | 1.94 | 4.2 | 3.7 | 0.5 | 290 | 0.00905 | 1.8 |
| 3.0 | 1.29 | 3.2 | 2.8 | 0.4 | 287 | 0.00600 | 1.5 |
| 1.5 | 0.645 | 2.1 | 2.0 | 0.1 | 283 | 0.00298 | 3.0 |

# LIST OF SOURCES CITED

The following abbreviations of the names of periodicals are used:

Am. J. Phys.—American Journal of Physics
Ann. Physik—Annalen der Physik
Ann. Physik u. Chem.—Annalen der Physik und Chemie
Bull. Chem. Soc. Japan—Bulletin of the Chemical Society of Japan
Can. J. Research—Canadian Journal of Research
Chem. and Met. Eng.—Chemical and Metallurgical Engineering
Chem. Eng. Progress—Chemical Engineering Progress
Chem. Revs.—Chemical Reviews
Compt. rend.—Comptes rendus hebdomadaires des séances de l'académie des sciences
Forsch. Gebiete Ingenieurw.—Forschung auf den Gebiete des Ingenieurwesens
Helv. Phys. Acta—Helvetica Physica Acta
Ind. Eng. Chem.—Industrial and Engineering Chemistry
Ing.-Arch.—Ingenieur-Archiv
J. Am. Chem. Soc.—Journal of the American Chemical Society
J. Chem. Phys.—Journal of Chemical Physics
J. Ind. Eng. Chem.—Journal of Industrial and Engineering Chemistry
J. Math. Phys.—Journal of Mathematics and Physics
J. Phys. and Colloid Chem.—Journal of Physical and Colloid Chemistry
Nature—Nature
Naturwiss.—Naturwissenschaften
Phil. Mag.—Philosophical Magazine and Journal of Science
Phys. Rev.—Physical Review
Physik. Z. Sowjetunion—Physikalische Zeitschrift der Sowjetunion
Proc. Roy. Soc. (London)—Proceedings of the Royal Society (London)
Sitzber. Akad. Wiss. Wien—Sitzungsberichte der Akademie der Wissenschaften in Wien
Science Progress—Science Progress
Trans. Am. Inst. Chem. Engrs.—Transactions of the American Institute of Chemical Engineers
Trans. Am. Inst. Elec. Engrs.—Transactions of the American Institute of Electrical Engineers
Trans. Am. Soc. Mech. Engrs.—Transactions of the American Society of Mechanical Engineers
Trans. Faraday Soc.—Transactions of the Faraday Society
Trans. Roy. Soc. (London)—Philosophical Transactions of the Royal Society (London)
Z. angew. Math. u. Mech.—Zeitschrift fur angewandte Mathematik und Mechanik
Z. Physik—Zeitschrift fur Physik
Z. physik. Chem.—Zeitschrift fur physikalische Chemie
Z. tech. Physik—Zeitschrift fur technische Physik
Z. Ver. deut. Ing.—Zeitschrift des Vereines deutsche Ingenieure

# LIST OF SOURCES

1. Ackermann, G., Ing.-Arch., 5, 124-146 (1934).
2. Adam, N. K. The Physics and Chemistry of Surfaces. 3d ed., London, Oxford University Press, 1941.
3. Alty, T., Proc. Roy. Soc. (London), A131, 554-564 (1931).
4. ——, Nature, 130, 167-168 (1932).
5. ——, Phil. Mag., 15, 82-103 (1933).
6. ——, Science Progress, 31, 436-448 (1937).
7. ——, Nature, 139, 374 (1937).
8. Alty, T., and C. A. McKay, Proc. Roy. Soc. (London), A149, 104-116 (1935).
9. Alty, T., and F. H. Nicoll, Can. J. Research, 4, 547-558 (1931).
10. Baule, B., Ann. Physik, 44, 145-176 (1914).
11. Bennewitz, K., Ann. Physik, 59, 193-224 (1919).
12. Berg, L., and D. O. Popovac, Chem. Eng. Progress, 45, 683-691 (1949).
13. Bikerman, J. J. Surface Chemistry for Industrial Research. New York, Academic Press, 1948.
14. Bosnjakovic, F., Forsch. Gebiete Ingenieurw., A3, 135-143 (1932).
15. Bowman, J. R., and R. C. Briant, Ind. Eng. Chem., 39, 745-751 (1947).
16. Boynton, W. P. Applications of the Kinetic Theory to Gases, Vapors, Pure Liquids, and the Theory of Solutions. New York, Macmillan, 1904.
17. Bradish, C. J., C. M. Brain, and A. S. MacFarlane, Nature, 159, 28-29 (1947).
18. Brookfield, K. J., H. D. N. Fitzpatrick, J. F. Jackson, J. B. Matthews, and E. A. Moelwyn-Hughes, Proc. Roy. Soc. (London), A190, 59-67 (1947).
19. Bruce, H. D., Proc. Roy. Soc. (London), A171, 411-421 (1939).
20. Carman, P. C., Trans. Faraday Soc., 44, 529-536 (1948).
21. Chapman, S., and Cowling, T. G. The Mathematical Theory of Non-Uniform Gases. London, Cambridge University Press, 1939.
22. Colburn, A. P., and O. A. Hougen, "Studies in Heat Transmission," Bulletin of the University of Wisconsin, Engineering Experiment Station, Series No. 70, October, 1930.
23. Crout, P. D., J. Math. Phys., 15, 1-54 (1936).
24. Eucken, A., Naturwiss., 25, 209-218 (1937).
25. Fick, A., Ann. Physik u. Chem. (Poggendorf), 94, 59-86 (1855).
26. Fowler, R. H. Statistical Mechanics. 2d ed., London, Cambridge University Press, 1936.
27. Fuchs, N., Physik. Z. Sowjetunion, 6.3, 224-243 (1934).
28. Furry, W. H., Am. J. Phys., 16, 63-78 (1948).

29. Giddings, H. A., and P. D. Crout, J. Math. Phys., *15*, 124-178 (1936).
30. Gilliland, E. R., and T. K. Sherwood, Ind. Eng. Chem., *26*, 516-523 (1934).
31. Hertz, H., Ann. Physik u. Chem. (Poggendorf), *17*, 193-200 (1882).
32. Hertzfeld, K. F., J. Chem. Phys., *3*, 319-323 (1935).
33. ———, Lehrbuch der Physik, ed. Müller-Pouillet. Braunschweig, Vieweg, 1925. Bd. III/2, pp. 229-237.
34. ———, Chem. Revs., *34*, 51-106 (1944).
35. Higbie, R., Trans. Am. Inst. Chem. Engrs., *31*, 365-389 (1935).
36. Jahnke, E., and F. Emde. Tables of Functions. New York, Dover Publications, 1943.
37. Jones, H. A., I. Langmuir, and G. M. J. McKay, Phys. Rev., *30*, 201-214 (1927).
38. Keenan, J. H. Thermodynamics. New York, Wiley, 1941.
39. Kennard, E. H. Kinetic Theory of Gases. 1st ed., New York, McGraw-Hill, 1938.
40. Kirschbaum, E., Forsch. Gebiete Ingenieurw., *B4*, 146-149 (1933).
41. Kleeman, R. D. A Kinetic Theory of Gases and Liquids. 1st ed., New York, Wiley, 1920.
42. Knoblauch, O., and Reiher, H., Z. tech. Physik, *4*, 432-433 (1923).
43. Knudsen, M., Ann. Physik, *29*, 179-193 (1909).
44. ———, Ann. Physik, *34*, 593-656 (1911).
45. ———, Ann. Physik, *47*, 697-708 (1915).
46. Kundt, A., and E. Warburg, Ann. Physik u. Chem. (Poggendorf), *156*, 177-211 (1875).
47. Langmuir, I., Phys. Rev., *2*, 329-342 (1913).
48. ———, J. Am. Chem. Soc., *37*, 417-458 (1915).
49. ———, Phys. Rev., *8*, 149-176 (1916).
50. ———, J. Am. Chem. Soc., *54*, 2798-2832 (1932).
51. LeBlanc, M., and G. Wuppermann, Z. physik. Chem., *91*, 143-154 (1916).
52. Lennard-Jones, J. E., and A. F. Devonshire, Proc. Roy. Soc. (London), *A156*, 6-36 (1936).
53. Lewis, W. K., J. Ind. Eng. Chem., *8*, 825-833 (1916).
54. Little, E. M., Phys. Rev., *52*, 255-256 (1937).
55. Loeb, L. B. The Kinetic Theory of Gases. 2d ed., New York, McGraw-Hill, 1934.
56. Mache, H., Sitzber. Akad. Wiss. Wien., *119*, 1399-1423 (1910).
57. ———, Z. Physik, *107*, 310-321 (1937).
58. ———, Z. Physik, *110*, 189-196 (1938).
59. Marcelin, R., Compt. rend., *158*, 1419-1421 (1914).

## LIST OF SOURCES

60. ———, Compt. rend., *158*, 1674-1676 (1914).
61. Maxwell, J. C., Trans. Roy. Soc. (London), *170*, 231-256 (1879).
62. Melville, H. W., Trans. Faraday Soc., *32*, 1017-1020 (1936).
63. Millikan, R. A., Phys. Rev., *21*, 217-238 (1923).
64. Miyamoto, S., Bull. Chem. Soc. Japan, 7, 8-17 (1932).
65. ———, Trans. Faraday Soc., *29*, 794 (1933).
66. Möbius, W., Z. tech. Physik, 6, 58-60 (1925).
67. Nusselt, W., Z. angew. Math. u. Mech., *10*, 105-121 (1930).
68. Penner, S. S., J. Phys. and Colloid Chem., *52*, 367-373 (1948).
69. ———, J. Phys. and Colloid Chem., *52*, 949-954, 1262-1263 (1948).
70. Pollitzer, F., Z. tech. Physik, *4*, 433-434 (1923).
71. Prüger, W., Z. Physik, *115*, 202-244 (1940).
72. ———, Sitzber. Akad. Wiss. Wien, Math.-naturwiss. Kl., Abt. IIa, *149*, 31-58 (1940).
73. Raman, C. V., and L. A. Ramdas, Phil. Mag., *3*, 220-223 (1927).
74. Rayleigh, Lord, Phil. Mag., *33*, 1-18 (1892).
75. Reynolds, O., Proc. Roy. Soc. (London), *22*, 401-407 (1874).
76. ———, Trans. Roy. Soc. (London), *166*, 725-735 (1876).
77. ———, Trans. Roy. Soc. (London), *170*, 727-845 (1879).
78. Sauerwald, F. Lehrbuch der Physik, ed. Müller-Pouillet. Braunschweig, Vieweg, 1926. Bd. III/12.
79. Schirmer, R., Z. Ver. deut. Ing., No. 6, 170-177 (1938).
80. Schreber, K., Z. tech. Physik, *4*, 19-27 (1923).
81. ———, Z. tech. Physik, *4*, 434-436 (1923).
82. ———, Z. tech. Physik, *14*, 81-85 (1933).
83. Slepian, J., and W. W. Brubaker, Phys. Rev., *55*. 1147 (1939).
84. ———, Phys. Rev., *57*, 250 (1940).
85. ———, Trans. Am. Inst. Elec. Engrs., *59*, 381-384 (1940).
86. Smoluchowski, M., Ann. Physik u. Chem. (Wiedmann), *64*, 101-130 (1898).
87. Stefan, M. J., Sitzber. Akad. Wiss. Wien, Math.-naturwiss. Kl., Abt. II, *68*, 385-423 (1873).
88. Thomson, W., Phil. Mag., *42*, 448-452 (1871).
89. Tolman, R. C. Statistical Mechanics with Applications to Physics and Chemistry. ACS Monograph Series, New York, Chemical Catalog Co., 1927.
90. Tschudin, K., Helv. Phys. Acta, *19*, 91-102 (1946).
91. Tucker, W. H., and T. K. Sherwood, Ind. Eng. Chem., *40*, 832-838 (1948).
92. Uhara, I., Bull. Chem. Soc. Japan, *18*, 412-427 (1943).
93. van Wijk, W. R., Z. Physik, *75*, 584-598 (1932).
94. Volmer, M. Kinetik der Phasenbildung. Dresden, Steinkopff, 1939.

95. Volmer, M., and I. Estermann, Z. Physik, 7, 1-12 (1921).
96. ——, Z. Physik, 7, 13-17 (1921).
97. Wallauchek, E., Dipl. Ing., Lehrkanzel für Experimentalphysik der technischen Hochschule Wien; personal communication, April 25, 1950.
98. Whitman, W. G., Chem. and Met. Eng., 29, 146-148 (1923).
99. Wiedmann, M. L., and P. R. Trumpler, Trans. Am. Soc. Mech. Engrs., 68, 57-64 (1946).
100. Winkelmann, A., Ann. Physik, 22, 1-31, 152-161 (1884).
101. Wyllie, G., Proc. Roy. Soc. (London), A197, 338-395 (1949).

Bei Fragen zur Produktsicherheit wenden Sie sich bitte an:
If you have any questions regarding product safety,
please contact:

Walter de Gruyter GmbH
Genthiner Straße 13
10785 Berlin
productsafety@degruyterbrill.com